P9-CMC-623
DISCARD

# Service Industries and Economic Development

## Case Studies in Technology Transfer

*Ronald K. Shelp*
*John C. Stephenson*
*Nancy Sherwood Truitt*
*Bernard Wasow*

Sponsored and Administered by the FMME
Fund for Multinational Management Education

PRAEGER

PRAEGER SPECIAL STUDIES • PRAEGER SCIENTIFIC

New York • Philadelphia • Eastbourne, UK
Toronto • Hong Kong • Tokyo • Sydney

**Library of Congress Cataloging in Publication Data**
Main entry under title:

Service industries and economic development.

Includes bibliographic references and index.
1. Service industries—Developing countries—Case studies. 2. Technology transfer—Developing countries—Case studies. I. Shelp, Ronald Kent.
HD9989.D44S47 1984 338.4'7'00091724 84-8304
ISBN 0-03-070678-5 (alk. paper)

Published in 1984 by Praeger Publishers
CBS Educational and Professional Publishing,
a Division of CBS Inc.
521 Fifth Avenue, New York, NY 10175 USA



456789 052 987654532

Printed in the United States of America
on acid-free paper

*IN MEMORIUM*

*Joseph D. Peno, a pioneer in conceptualizing technology in the insurance industry, who died in a tragic accident before he could see his ideas come to fruition.*

*John C. Stephenson, whose efforts and insightful perceptions contribute greatly to this book, who died shortly before its publication.*

# Contents

# List of Figures

CHAPTER 1

# The Role of Service Technology in Development

*Ronald Kent Shelp*

*Agriculture, mining, and manufacturing are the bricks of economic development. The mortar that binds them together is the service industry.*

This book examines the technologies of service industries and how they are transferred to developing countries. Services are the just-off-the-boat immigrants in the eyes of most economic theorists, and as such have been more ignored than understood. Service technologies made their appearance on the scene long after everyone had been taught that wealth was created only by the production of tangibles such as agricultural crops, mined minerals, manufactures, and erected structures. Consequently, the provision of services has not been examined adequately. In fact, association of services with technology comes as a surprise to many. To help fill that gap, this book presents three case histories of service technology transfer to developing countries by multinational companies.

These searching studies present hard evidence showing that services are not just an appendage of mainstream economic activities, but rather their multiplier. Service technologies should be of special interest to the developing countries because they are in the rare and enviable position of being a subject of little contention in the North/South dispute about technology transfer. Much of service technology is transferred as a matter of course. The extent and rapidity of transfer depend on host country attitudes. The productivity of the developing economies receives a boost when available advanced technologies are put to use. Conversely, growth is retarded when the latest in service industry developments is not put to use. Thus, services can play a much larger role in meeting legitimate expectations in developing countries than has so far occurred.

## THE BACKGROUND

The role of services in the economy was established in the public mind in the preindustrial era. Personal services, such as those provided by housekeepers, barbers, and plumbers, were deemed typical of the service industry as a whole, along with education and other community services furnished by the public sector. Since, by definition, personal services involve a one-to-one relationship between the service provider and the service receiver, their productivity generally cannot be readily increased by capital and technology. In consequence, as the onset of industrialization made more remunerative alternatives available, personal service jobs in comparison came to connote low pay, and soon their share of total employment began to dwindle.

However, data available on the developed countries indicate that in the industrial and postindustrial stages, other service jobs have been slowly but consistently increasing their share of the labor force. In the United States, for example, between 1960 and 1981 employment in services increased by an average 3.2 percent annually, while employment in manufacturing grew at a rate of 1.0 percent per year.[1]

In April 1982, the work force employed in nongovernment U.S. service industries for the first time exceeded the number of workers employed by the entire goods-producing sector (manufacturing, construction, mining).[2] Today, if we include government, approximately seven of ten Americans are employed in the service sector. The contribution of services to the U.S. GNP has progressed at a similar rate. Some two-thirds of the GNP results from service production.[3]

There is an explanation for the apparent contradiction whereby industrialization initially led to a decline in service jobs but subsequently brought about a surge in service job creation. While personal service jobs were declining, industrialization called forth a whole range of new services. Some of these were the result of newfound affluence, as more and more people could afford more and better health care, education, amusement, and recreation. Other services were needed to increase the productivity of production—wholesale trade, information processing, financial services, communications. These services and others like those examined in the case studies—engineering consulting, retailing, and insurance—became highly productive when modern technology supplied them with computers, satellite and other rapid communications, and systems analysis.

Thus, service jobs moved away from the low end of the economic spectrum toward the other extreme. Much of the service-oriented job growth in advanced nations has taken place in professional, managerial, administrative, and problem-solving categories. Increasingly, education became the name-of-the-game in service jobs. From 1972 to 1981, employment in the business service category increased by 82 percent in the United States. This growth was led by highly skilled, highly paid professions. Legal services grew by 96 percent, followed by engineering and architectural (68 percent), accounting (64 percent), and financial services (36 percent). During this time, manufacturing jobs increased by only 5 percent and construction by 7 percent.[4] At the same time, wholesale and retail commerce progressed from basic warehousing and storekeeping to sophisticated product marketing.

## CLASSIFICATION OF SERVICES

Keeping in mind the gradual change in the morphology of services used by developed countries as they passed through various stages of development will help us understand the differing roles that service industries play in economies that are now in earlier stages of development. Five categories of services that basically parallel the phases of economic change may be usefully discerned.

### Unskilled Personal Services

Historically, housekeeping service for females, military conscription for males, and street vending for both sexes have been the main sources of service activity in traditional societies. These kinds of jobs provided entry opportunities for excess population to become socialized into urban life.

### Skilled Personal Services

As productivity increased in agriculture-based societies so that production exceeded subsistence requirements and the industrial revolution began, opportunities opened up for skilled artisans and shopkeepers in wholesale and retail commerce, repair and maintenance, and clerical employment. In addition, proliferating government services supplying infrastructure and social services to the new industries and the burgeoning urban population provided service employment opportunities.

### Industrial Services

As industries competed in the marketplace, they grew in sophistication and required more highly skilled services. These could be more efficiently provided by specialized service organizations, such as legal and accounting firms, banks and insurance companies, real estate brokers, and import/export trading companies. Vastly expanded transport and communication requirements were met in most of the world by public sector services.

### Mass Consumer Services

Productive industries spread wealth among the population. The expanding discretionary purchasing power in turn gave rise to creative consumer service industries able to enjoy economies of scale while accommodating vastly growing consumer demand for travel (airlines, hotels, auto rental), culinary enjoyment and convenience (restaurant and fast food chains), entertainment (films, television, clubs), and a plethora of inventive personal services ranging from videogame parlors to encounter groups. The greatest growth was registered by health care services, based on advances in medical research and the ability of patients to pay directly or through insurance programs. In all of these service industries, the affluent mass markets made efficiency possible.

### High Technology Business Services

The introduction of computers, microchips, lasers, earth satellites, and technology associated with nuclear and biological breakthroughs brought about a revolution in services to business. Automation, electronic data processing (EDP), robotics, nuclear power, hydroponics, and other innovations were seized upon by a new generation of engineering and management consulting firms, think tanks and university-affiliated research centers, and software programmers to provide state-of-the-art technical services to industry.

## SERVICES AND THE DEVELOPING COUNTRIES

There is evidence that developing countries do not neatly climb this five-stage ladder, following in the footsteps of the West. That option is not really available to them even if they wanted to take it. For one thing, there is the demonstration effect. Developing country consumers, made aware by the communication media of the services available on the world market, do not readily accept the notion that the availability of modern health, education, leisure, and other services will have to await their arrival at the appropriate stage of development. This increased consumer awareness in the developing nations is heightened by the urbanization process. Whereas in the West urbanization came about as a result of industrialization, in many developing countries it has run ahead of industrial development. It has intensified the demand for a wide array of services ranging from basic infrastructure to general welfare to financial and commercial services.

The development strategies followed by developing countries comprise another important factor responsible for this unique pattern of service development. Their manufacturing and agricultural sectors recognize that they have to compete in world markets if they are to progress. The service technologies available to them from within might be appropriate for the goal of creating jobs and reducing imports, but they are likely to be uncompetitive. To succeed in export markets, they must employ the best services available anywhere; otherwise they will be out of the game in today's complex international economy.

In their drive to catch up with the developed countries, developing countries leapfrog stages in the progression of their service industries. Skilled personal services increase along a broad front, and at the same time industrial services spring to life where there is demand for them. In export-oriented enclaves, usually in industrial belts around the major cities and free ports, high technology business services are available. In the capital cities, mass consumer services thrive. To meet these needs, many developing countries, instead of trying to develop their own service technologies, avail themselves of opportunities to obtain service technology transfer from developed countries and apply it where it is needed within their economies, in islands of modern productivity.

The importance of services as facilitators of economic productivity may be illustrated by the example of agriculture—a sector of crucial importance to nearly all developing countries. Some two-thirds of the world's arable land is operated by farmers of limited resources (subsistence and other family-operated farms relying entirely on hand labor). For decades, governments and international organizations have been attempting to raise the productivity of these farmers who represent some three-quarters of the world's farm population. Toward that end, new technologies have been offered (improved seed, fertilizer, and pesticides; mechanization and other means of rationalizing cultivation). A major constraint has been the unwillingness of these farmers of limited resources to adopt these new technologies.

More intensive exploitation requires the investment of money as well as labor. Whereas easy credit may be available as a matter of public policy, typical developing country farmers have no reserves that would enable them to absorb the frequent weather-related adversities affecting their harvest. Hence, they are reluctant to risk the welfare and perhaps the survival of their families by borrowing to finance a risky investment. They tend to prefer to stay with the tried-and-true traditional techniques learned from their ancestors, even if that means accepting lower yields. The wide availability of crop insurance at reasonable cost would provide an incentive to undertake the risk taking necessary to introduce new agricultural methods and technologies. The introduction of this service product, crop insurance, into developing nations could dramatically accelerate their agriculural development. Additional millions of tons of foodstuffs could be produced where they are needed the most—on the family farm.

The role of services in the economies of developing countries is already more significant than is commonly realized. An Organization for Economic Cooperation and Development (OECD) study in the early 1970s of 56 developing countries revealed that some 28 percent of their labor force was employed in services.[5] For more than one-third of the countries, service employment accounted for 30 percent or more of the total. Several had very high proportions of service employment, and three, in fact, qualified as service economies, since over half the labor force was employed in services. Some countries virtually skipped the industrial phase altogether and became tertiary or service economies before they were full-fledged industrial societies.

Developing economies in which services constitute a major portion of economic output are much more numerous than those in which services are a major provider of employment. In 19 of the countries studied, close to one-third of the sample, services were producing more than one-half the gross domestic product (GDP). So, as a general statement, services contribute somewhat more to production than they do to employment.

But the important question is which service activities are yielding these dramatic percentages of jobs and production. The assumption is that a high proportion of service employment in developing economies would be found in government and in personal services—vendors, lottery sellers, shoeshiners and the like. And the importance of these kinds of services cannot be denied. They provide over one-half the service sector employment in the countries studied. But business and other more sophisticated services play a surprisingly important role. For example, commerce, which includes banking, insurance, real estate, and wholesale/retail trade, accounts for nearly 30 percent of service employment. Transportation, communications, and storage account for 15 percent. Retail trade contributes about the same percentage to the GNP—30 percent—that it does to employment. Other services, including financial, account for 20 percent of the GNP, and transport, storage, and communication contribute 6 percent. In short, these kinds of services—not personal services—make up the majority of service contribution to production.

Finally, the role services play in the international trade of developing countries is equally surprising. In a study of the invisibles transactions of 18 developing countries from all regions covering the period 1967–73,[6] although trade in goods was found to be more significant, in a few of the countries service exports grew dramatically faster than goods exports. Of special interest was how important service imports are to these economies. In the majority, the import of services tends to be increasing at a rate that matches or exceeds the import of goods. This suggests that services are much more important to growth patterns than we traditionally think.

At the policy level, the perception of many developing countries of the importance of the service sector to their economies runs ahead of their industrial country counterparts who are still prisoner to the traditional concept that production of wealth is synonymous with production of goods. The OECD study on services mentioned earlier brought out the startling fact that in 28 developing countries, 56 percent of all fixed capital formation took place in the service sector (one-half of this investment being related to transport, storage, communications, and the construction of dwellings).[7]

Recognition of the importance of building a national service sector, however, is more often a reflection of a quest for national independence than a central part of an economic strategy. That means that considerable economic costs are borne in order to allow local services to substitute for imported service technologies or foreign service investments. Yet advanced service technologies could often be transferred from abroad without incurring the expense of creating them from the ground up.

## THE TRANSFER OF TECHNOLOGY DEBATE

The availability of foreign technologies to developing countries is a subject of controversy. In United Nations fora, in particular, the issue of technology transfer has been discussed in passionate dialogues between the North and the South. Much of the time, consensus founders on the rocks of the underlying incompatibility between the two perspectives. The developing nations need technology transfer to modernize their economies and compete in world markets. They cannot afford to emulate the West and build up their technical capabilities over a century or more. To catch up, they demand the transfer of Western technologies as a matter of social justice. In developed countries, however, most of the technical know-how is not in the public domain, but is owned by private companies that have spent huge sums to develop their technologies. If they are willing to transfer it at all, they will do so only for what they consider an equitable return on their investment.

For the most part, however, this debate has passed service industries by. The focus has been on what is perceived as "high technology," which translates in most minds to industrial and manufacturing production—not production of services. Although economists and policymakers of developing countries may recognize the importance of services, they are no more likely to identify services with technology than are their industrial nation counterparts. So the debate itself has ignored services. Besides, many of the issues that have been the center of focus—proprietary rights, patents, trademarks, protection of processing—are not very relevant to most services. Yet the policies of many developing countries toward services, while not intended to affect the transfer of service technology, do exactly that. The case studies in this volume show that much of service technology is transferred by means of foreign investment. Policies that restrict foreign investment in services are, unwittingly, deterring the transfer of service technology.

## JUDGMENTS ON APPROPRIATE SERVICE TECHNOLOGY

Even those few who recognize the value of service technology will want to make judgments on the relative importance of different service activities. This will be especially true of those technologies to be imported. But the dividing line between "frivolous" and "productive" services is not easy to draw. The experience of the centrally planned economies of the Soviet bloc, based on the Leninist view that most services activities are nonproductive or parasitic, illustrates this dilemma.

In the Soviet scheme, a distinction is drawn between "productive services," which are defined as those directly related to physical production and include some trade in goods, transportation, and communication services used in the production of goods, and "nonmaterial services," such as passenger transport, other communications, housing, health, education, finance, insurance, public administration, and repair and personal services. The latter are viewed as contributing to the redistribution of income rather than the creation of output. Thus, development of most services is given a very low priority. The upshot of this policy has been that in the Eastern European countries, the proportion of the GNP produced by the service sector was stagnant between 1965 and 1975.[8]

What effect has this policy had on economic development? Available evidence suggests this underdeveloped service sector retards growth. For example, the relegation of wholesale commerce to the "unproductive" category in the Council for Mutual Economic Assistance of the Eastern European Nations (COMECON) countries has consistently produced shortages of goods, much to the dissatisfaction of the frustrated consumer. The managers of state industrial enterprises have found they cannot depend on a competitive array of service providers to furnish support services for a changing production schedule. A well-done case study on Hungary reaches a conclusion that could be applied generally: Neglect of the service sector probably hampers industrial growth.[9]

## THE PROCESS OF SERVICE TECHNOLOGY TRANSFER

In developing countries, the prevailing conception of technology is that of a black box that contains all there is to know about producing certain goods or services. The goal, then, is to get hold of that black box. In reality, the transfer of technology depends more on the capacity of the recipient to absorb it than on any other factor. Furthermore, the essential ingredient in any transfer of technology is time. It cannot be executed except by a sustained interaction between the transferor and the recipient over a long period of time. Transfer of technology is essentially a training process that transfers skills rather than information. Application of technology requires the knowledge of not only how it works, but also why it works.

When it comes to transferring service industry technology, understanding how it works is difficult, because reliable statistical data are hard to come by. The process does not lend itself easily to quantification. To gain an insight into the processes of transferring servicc technologies, the three studies presented in this book take the case history approach. They examine in considerable depth the experiences of three service companies engaged in the transfer of technology. Taken together, these case studies indicate there are consistent elements in the technology transfer process. The process requires the continuous, long-term interaction of people within a well-organized framework of accountability with high standards for technical performance.

### Methods

On-the-job *training* may be provided by expatriate and short-term experts; formal training may be organized in-country, at headquarters, or in third countries. The cases show that the three multinationals made a sustained effort to train national personnel. The range was from a few hours spent by Sears, Roebuck and Co. sales staff to over 1,200 hours invested by Sears executives to the years of training provided by Bechtel in transferring technology in the nuclear power industry.

*Documents* such as manuals, blueprints, and samples of model communications are used to effect the transfer of systematic technologies. The Sears case refers to various types of documents and shows how technology relating to the development of new product ideas was transferred.

Establishment and maintenance of quality *standards* through supervision are a major contribution of the transferor of service technology. The foreign firm is attractive to its clients in the developing countries in part because it enforces high standards, as documented in the American International Group (AIG) and Bechtel studies.

The structure of *reports* establishes the confidence that stems from the building of systemic knowledge; responses from the home office provide technical and managerial information. The AIG case illustrates the learning opportunities presented by rapid feedback telex communication with headquarters.

### Impact

*Performance capability* is created within the local affiliate as a result of formation of local personnel. That is precisely the intent of Bechtel and a way of life for AIG.

Much service technology is easily *diffused* because it is out in the open. Once a service technology has passed the experimental stage and has proven itself successful, it is quickly copied by local competitors. Local service industries adopt the newly introduced technologies and hire away trained personnel. The diffusion process is sustainable because the parent company stays in business not by jealously guarding its proprietary secrets, but by continually creating new technologies to be passed along.

Service technology transfer *creates jobs* by establishing a new capability to provide a marketable service, both in the local affiliate and among its competitors and secondarily among their subcontractors. All three studies illustrate the point. The Sears case, in particular, presents painstaking evidence of primary and secondary job creation.

The transfer of advanced technology increases the *productivity* of the work force employed in that industry. If that industry provides services to other industries, it raises their productivity also. Ability to provide insurance at a lower cost because of organizational efficiencies means that higher salaries can be paid to insurance company staff and that lower premiums can be charged to clients.

**Transferor Rationale**

Consulting companies may be contracted for the specific purpose of transferring service technologies. The Bechtel Organization is but one example of a large and growing number of consulting companies that make it their business to transfer technology in such fields as engineering, management, and agronomy.

A *long-term relationship* between the service technology supplier and receiver benefits both parties. Be it a trading or investment arrangement, a long-term affiliation can give the local company a leg up on its competitors and assure the parent company of a stable market. To maintain their competitiveness, service companies continually create new technologies through their own research and development efforts. For that reason, they can live with the eventual incidental loss to others of the technologies that they developed—provided that the payout has been sufficient to finance new research and development. When the probability of continuing the long-term relationship goes down, as in the case of Sears Peru, both the parent and local company lose.

**Institutional Framework**

The transfer of service technology occurs within an institutional framework. The need is for mechanisms to enforce standards for the transfer of experience over time.

Technology transfer is a person-to-person sharing of technical and managerial experience, attitudes, and viewpoints.

The more complex the technology, the more crucial is the favorable personal "chemistry" between transferor and recipient; both need to be highly motivated. A relationship of trust and good will between the host country and the country of the parent company, as well as between the affiliate and the home office, enhances the probability of successful transfer. Conversly, an environment fraught with suspicion and resentment, attempts at forcing career advancement of indigenous personnel by quotas, threats of punitive action, and disregard of objective standards of performance create a climate in which technology transfer is hindered.

### Conclusion

The studies in this book shed new light on service industries and their role in economic development. They make a strong case for service industries being a major instrument in speeding the process of development.

Irving Leveson of the Hudson Institute has pointed out, "It may no longer be necessary for countries to wait until they have reached the current U.S. level of GNP per capita in order to experience the kinds of changes which U.S. service industries are going through today. Service industry productivity changes associated with the application of modern management techniques and technologies are clearly subject to international transmission. With more detailed knowledge of the nature and sources of service industry productivity growth, we should be able to judge the degree to which service industries in other countries will enter a phase of more extensive innovation at earlier income levels than in the U.S."[10]

## ECONOMIES OF SCALE

It is commonly assumed that economies of scale do not apply to service industries. The three cases presented in this book argue the opposite, and suggest that the very structure of the industries means that significant efficiencies can be obtained when services are provided on a broad, multinational basis.

Some unskilled and skilled personal services, which directly relate the individual who provides the service with the person who receives it, may well be resistant to economies of scale. However, mass consumer, high technology business and other services owe their existence to mass markets. The essence of insurance, the spreading of risk, is based on the law of large numbers; transportation, both passenger and freight, becomes affordable through providing services in bulk; and the mere name "mass merchandising" outlines the selling strategy that gave birth to the industry.

The service technologies developed by Sears, AIG, and Bechtel could not have been created on a narrow market base. This fact alone suggests that many developing countries will need to be part of a high volume market if they are effectively to develop mass consumer and high technology business services.

As much as anything else, economies of scale in the service industry have to do with people. The Sears and AIG cases demonstrate the value of having available specialized experts who have a long-term relationship with the affiliate and come at comparatively low cost because they are supported by the parent's worldwide operations. That, and the ability demonstrated by Bechtel to send in whole teams of experts for limited periods of time, can never be matched by the availability of local personnel or by hiring short-term individual experts on the open consulting market. Consultants have the disadvantage of not operating within the same institutional network. This makes their familiarization time longer and their competence harder to evaluate.

## OPPORTUNITY COSTS

The educational requirements of modern service industries are significantly higher than those of manufacturing, agriculture, and extractive industries. In the United States, service industry employees have an average of over 13 years of education compared with fewer than 11 years for manufacturing. In developing countries, where the pool of educated labor is limited, consideration must be given to the most efficient use of this limited human resource.

Indeed, as Professors Wasow and Hill found in a study on the insurance industry, which would be applicable to many service industries, the skill levels of services like insurance raise important public policy questions.[11] Should skilled people and physical capital be tied up to provide the managers, buildings, and computers for all high skill service industries, or should these resources be directed into other sectors—machinery, fabricated metal, etc.? Should, in other words, import substitution take place in the service sector or in manufacturing industries? Wasow and Hill conclude that other sectors seem to offer alternatives where the skill and capital requirements are lower and where the utilization of limited resources can be expected to generate a great deal more value added and sometimes more jobs than in certain service sectors. This suggests that arrangements leading to a transfer of foreign service technology, rather than trying to develop a service industry alone, might be the best policy decision for many developing countries.

Perhaps the most compelling argument for importing service industry technology is the rapid boost it can give to the process of economic development. The development of new service technologies in the developed world is not just a function of the volume of business made possible by large markets. It is also a function of the sophisticated educational establishments of these countries and indeed of their entire social and political institutional framework. These underpinnings necessary for economic progress can be replicated in developing countries only over time. It would be tragic if their economic development were retarded by waiting for this transformation process to occur.

At a time when even the industrial countries are just beginning to evaluate the importance of service industries to their economies, these studies dramatize the potentially valuable role that modern service industries play in the economic development process of all kinds of economies.

## NOTES

1. *Monthly Labor Review*, U.S. Department of Labor, November, 1982, p. 65.

2. U.S. Department of Labor, Bureau of Labor Statistics, New York, 1982.

3. U.S. Department of Commerce, Bureau of Economic Analysis, Washington, 1980.

4. U.S. Department of Labor, Bureau of Labor Statistics.

5. "Service Activities in the Developing Countries," Organization for Economic Cooperation and Development, Paris, 1974.

6. Ronald Kent Shelp, *Beyond Industrialization: Ascendancy of the Global Service Economy* (New York: Praeger, 1981).

7. "Service Activities in the Developing Countries," OECD.

8. Thad P. Alton, "Comparative Structure and Growth of Economic Activity in Eastern Europe," in *East European Economics, Post-Helsinki.* U.S. Congress, Joint Economic Committee, Washington, 1977.

9. Gur Ofer, *The Service Sector in Soviet Economic Growth* (Cambridge, Mass.: Harvard University Press, 1973).

10. Irving Leveson, *Productivity in Services—Issues for Analysis* (Croton-on-Hudson: Hudson Institute, 1980).

11. Bernard Wasow and Raymond D. Hill, "Public Sector Involvement in the Insurance Industry. The Implications for Economic Development." Study prepared for the Center for Applied Economic Research, New York University, New York, 1980.

CHAPTER 2

# Technology Transfer in the Insurance Industry: American International Group in the Philippines, Nigeria, and Kenya

*Bernard Wasow*

## INTRODUCTION

Throughout the decade of the 1970s, debate raged over an issue designated "technology transfer." The name covered a multitude of concerns, most of which fell within the larger issue of control of multinational corporations (MNC). Initially, most attention was focused on increasing access by local firms in developing countries to patents held by MNCs. However, it soon became obvious that patents were but the tip of the iceberg; that technology involves the complex of technical and managerial expertise to operate a business. A patent without the technical skills to utilize it and the managerial and marketing backup to produce a product and market it profitably is useless.

Thus, the international discussion has moved from a narrow focus on gaining access to patent rights to a much broader one on what constitutes modern technology, how it is transferred, and the benefits of that transfer. It is from this broader perspective that this study addresses technology transfer in insurance.

Few have thought of insurance as involving technology; yet technical and managerial decisions are at the heart of the business. What is technology in the insurance industry? How is it transferred? What is the role of the MNC in that transfer? What are some of the consequences of that transfer in the Third World countries in which it occurs? This case examines these questions through a study of the operations of the American International Group (AIG) in three countries: Kenya, Nigeria, and the Philippines. These countries were selected because in them AIG is involved with different types of insurance companies (life and general) and has a variety of relationships, including majority and minority interests (with private and with government partners). In addition, in these countries one finds companies with varying periods of involvement with AIG and, of course, geographic diversity.

This chapter first introduces the MNC responsible for technology transfers: AIG. It then discusses what technology is in the insurance industry and describes the mechanisms by which transfers take place and the impact of such transfers on the host country. Finally, it considers the MNC as a source of technology.

## THE AMERICAN INTERNATIONAL GROUP

AIG is an aggressively run, rapidly expanding insurance group. Started by C. V. Starr in Shanghai in 1919, it has differed from other U.S. insurance organizations from the outset in being internationally oriented. The organization is complex, consisting of a large number of partially or wholly owned agency companies, subsidiaries, reinsurance companies, and specialty companies in over 100 countries around the world.

Since 1969, the president of the company, Maurice R. Greenberg, has steadily consolidated and coordinated the different companies and subsidiaries under the AIG umbrella, instituting similar procedures for accountability and control throughout the corporate structure. The growth of business over the last decade has been impressive. Since 1969 net premiums earned in general insurance have grown over 17 percent annually, increasing to in excess of $2.25 billion. Premium income in life insurance too has increased by 18 percent per year. Investment income has increased more rapidly at 23 percent per year. These rapid growth rates have been sustained in large part by the growth of foreign business. As Table 2.1 shows, by 1982 47 percent of gross premium income, 37 percent of net premiums written, and 51 percent of operating income came from foreign operations in general insurance. In life insurance the figures are higher: Over 70 percent of AIG's life insurance operations are undertaken overseas, with the Philippines and other developing countries (as well as Japan) playing a predominant role.

In the Philippines, AIG is affiliated with seven companies. One of these, the Philippines American Life Insurance Company, is the subsidiary of the American Life Insurance Company, which in turn is a wholly owned subsidiary of AIG.

In the Philippines, AIG is represented by eight wholly owned companies and four affiliates. Among the former, The Philippine American Life Insurance Company is one of the best-known life insurers in the country, employing approximately 1,000 people; American International Underwriters (Philippines) Inc. operates independently of the Philippines American Life Insurance Company on the general insurance side. Management and staff in both organizations are almost exclusively Filipino.

In Nigeria, AIG now has a minority interest; the government of Nigeria required all foreign-owned insurers to sell 49 percent of their shares to the Federal Ministry of Finance in 1975. An additional 11 percent was required to be sold to three of the state governments in 1978. This is in accordance with a Nigerian law requiring that most types of business in Nigeria be at least 60 percent Nigerian owned. However, AIG has the right to name the managing director of the company, and plays a leading role in day-to-day decision making.

In Kenya, American Life Insurance Company has a life insurance branch and a newer general insurance branch, which until 1983 were located in different offices. In 1983, however, the company completed construction of a new building in Nairobi, and the two sections of the company were united under one roof.

**TABLE 2.1. American International Group, Inc. 1982**

| | *Value (millions of dollars)* | *Percentage Foreign* | *1972–82 Growth Rate* |
|---|---|---|---|
| General insurance | | | |
| Gross premiums written | 4,668 | 46.6 | 15.6[a] |
| Net premiums written | 2,263 | 36.7 | 14.5[a] |
| Net investment income | 302 | 39.7 | 27.4[a] |
| Operating income | 336 | 50.9 | 19.5[a] |
| Life insurance | | | |
| Premiums written | 860 | 71.4 | 21.0 |
| Total assets | 9,120 | NA[b] | 19.6 |

[a]Growth rate 1976–1982. Earlier data not comparable.
[b]NA, not applicable.
*Source*: AIG *Annual Report* 1982.

## THE TECHNOLOGY OF INSURANCE

Technology has been thought of, typically, as consisting of embodied "things": equipment, patents, machinery, circuits, new devices. Certainly, this is the description that has been asserted in much of the literature on the transfer of technology. However, economists also think of technology in a more general sense: Technology is the way in which inputs are converted into outputs. The insurance industry reduces the costs of bearing uncertainty. Through risk pooling and sometimes through risk management, the insurance company is able to reduce the expected loss and its variability from contingent events. The major input used in this production process is skilled people. The major problem solved by technology is how to use people and information efficiently.

Insurance technology might be classified into three broad areas. First, there are the technical tasks: underwriting, claim adjustment, finance, risk management, electronic data processing, product development, and reinsurance. Second, there are the marketing aspects, including agency management, which are important particularly to life insurance. Third, there are general administration and organization for a large service organization, the performance of which must constantly be monitored and controlled.

## TECHNICAL TASKS

*Underwriting*, which is perhaps the central function in insurance, involves the identification of risks and the assessment of their value, that is, the price that must be charged to assume them. Since a great many risks are rather routine, occurring in slightly varied form again and again, an experienced, educated person with a rating manual can in a few months learn how to handle these cases. On the other hand, there are also many sophisticated risks involving large factories, pipelines, offshore oil rigs, personal and product liability, and other lines, in which a great deal of technical and engineering knowledge must be brought to bear before a reasonable rate can be set. Engineers, lawyers, physicians, and other technicians must classify risks, which must then be analyzed by actuaries for the statistical properties of losses and claims. Where risks are small and often repeated, the statistical analysis of an actuary may provide the basic information necessary for rate setting. In many cases, however, risks are large or unusual enough to require substantial judgment in underwriting. In such cases, broad experience and keen discernment are essential. Indeed, even in "routine" cases, volumes of data are gathered and analyzed to assess the historical experience. The broader the experience of the individual and the institution involved in underwriting, the less likely will be unexpected losses, though some risk will always remain.

As Table 2.2 implies, underwriting decisions in relatively simple areas, such as automobile insurance, are already undertaken almost completely within the Third World countries in which AIG operates. On the other hand, more sophisticated underwriting, such as in aviation insurance or construction all-risk, is undertaken with a great deal of consultation with the home office. The range of these numbers reflects the varied extent of reliance on outside skills and decision making; it

**TABLE 2.2. Location of Underwriting Decisions in AIG Nigeria and Kenya: General Insurance**

| | *Percentage of New Premium Income Written without Consultation outside Country* | |
|---|---|---|
| *Line* | *Nigeria* | *Kenya* |
| Marine | 60 | 70 |
| Auto | 100 | 100 |
| Casualty | 90 | 50 |
| Accident and health | 98 | 70[a] |
| Property | 60[b] | 20 |

[a]Ordinary 100%, aviation 0%.
[b]Fire 100%, construction all-risk 0%.
*Source*: Interviews with AIG personnel and evaluation of 1978 company data.

indicates the extent of technical skills imported from underwriters elsewhere in the multinational.

In addition to the principal underwriting functions of assessing risks and valuing them, there is also an underwriting problem of forming a portfolio of risks. Even if individual risks appear to be priced correctly, the insurance company still needs to ask whether risks are independent of each other. For example, if adjacent buildings are insured against fire, then one fire could easily lead to two claims. Similarly, in areas subject to earthquake, hurricane, or other natural disaster, a large number of claims could be filed simultaneously if the underwriting of those risks were not coordinated. Insurance companies endeavor to keep claims relatively small and independent of each other. By limiting exposure by line, by geographic area, and by individual risk, they are able to reduce the variance of total losses, which is the central purpose of insurance. Reinsurance is a major tool in this effort to form a desirable portfolio of risks. Through reinsurance, exposure is reduced because risks are divided and spread throughout the world.

As with underwriting, the *settlement of claims* involves a range of technical skills. Complicated claims call for engineering skills to assess their legitimacy. Legal expertise is necessary in the decision to adjudicate or to settle claims out of court. The smooth handling of claims also involves a good deal of customer relations. In the Philippines, for example, the Philippine American Life Insurance Company has adopted policies introduced from the American Life Insurance Company of Japan in which only clearly fraudulent claims are adjudicated; honest mistakes are reunderwritten with a compromise. It has turned out that this system has earned the company much good will at moderate cost. Such a policy, which involves in this case a transfer of technology within AIG from Japan to the Philippines, entails decisions both technical—"How much will this cost us?"—and administrative—"How should we best deal with our customers?"

The third area of insurance involving considerable technical expertise is *finance*. Insurance companies accumulate large reserves against expected future payments and hold surplus and capital to deal with infrequent, very large losses. The management of these financial portfolios calls for technical evaluation of expected returns, riskiness, and the relationship between returns from various assets. Although important financial decisions are in many cases reserved for the home office—those dealing with currency exposure and the general structure of portfolios—specific investment decisions within the portfolio still depend on local expertise. Which mortgages to acquire, which government debt to purchase still must be determined, and local skills are essential for these decisions. In addition, the general accounting job, keeping track of the flow of funds within a large insurance company, is a demanding task since there is a large cash flow, much of which involves future responsibilities.

*Risk Management* involves the implementation of loss financing and loss control techniques to treat customers' exposure to loss. For example, the risk of a major fire can be reduced by requiring metal doors and sprinkling systems. In Nigeria, AIG's Nigerian deputy general manager for nonlife insurance actively pursued loss prevention measures in automobile lines, including the requirement that antitheft devices be placed on all cars insured by AIG. He also encouraged fleet managers to institute programs to provide incentives for drivers with good records. Careful underwriting often identifies high risk cases, which leads to loss management efforts. In loss financing, premiums can be reduced as the insured retains more of its own risk.

## MARKETING

The selling activities in an insurance company, already mentioned tangentially in discussing claims, are most important in the life insurance branch. Group life insurance and pension plans affect large groups of people and involve less selling than other types of insurance. These areas of life insurance are growing. Nevertheless, a great portion of life insurance business comes from small, individual contracts that are sold by agents. Programs for effective selling, for identifying potential clients, approaching them, and improving their persistence in paying premiums and avoiding lapses all are tasks that are important in life insurance. Fully one-third of all employees in the U.S. insurance industry are involved in sales. The proportion in the Philippine American Life Insurance Company is comparable (27 percent). Most of the salespeople do not work out of the head office; in fact, they seldom appear at headquarters, but rather are in the field making contact with clients and potential clients. The organization of such a sales force and the accountability and record keeping involved comprise a formidable administrative problem.

## ADMINISTRATION AND ORGANIZATION

In general, administration and organization not only of the sales efforts but of the entire insurance operation are a central aspect of insurance technology. When one is dealing with a large number of employees whose rate of work and accountability are not governed directly by equipment or machinery, then control systems, administrative chains, and hierarchies of command are essential for the smooth functioning of the business. Accounting, auditing, and reporting systems are all part of administration and organization. It is in this area, perhaps more than in any other, that considerable scope for innovation exists within insurance companies.

A central problem in organizing a large, dispersed service corporation is to determine where responsibility rests for each function and to set up a system of control and accountability to ensure that corporate goals are being met. Within AIG a number of functions are reserved for the home office, including decisions concerning the structure of the financial portfolio, currency exposure, and foreign exchange management and reinsurance policy. These decisions are inherently mutually interdependent. In underwriting, claim settlement, and marketing, in contrast, considerable responsibility and autonomy are vested in the local operations. Each subsidiary or affiliate has authority to underwrite risks and to handle claims line by line up to preassigned limits. These limits are adjusted when the home office or the regional decision makers believe that an increase or decrease in local independence seems warranted.

Decisions concerning the authority of the local affiliates are made on the basis of their performance in meeting corporate goals in underwriting results and premium growth. Where underwriting has been done skillfully and regional or home office officials are convinced that the skill for greater authority exists, the limits will be raised, as they were, for example, in the aviation line recently in the Philippine American General Insurance Corporation. On the other hand, where results appear poor and there is doubt that the underwriting function is being performed adequately, authority can be reduced, as it was in Kenya for workmen's compensation.

Given authority and given targets for production, a line manager in every subsidiary is responsible for the achievement of these targets, for the acquisition of good, new business. The manager of each subsidiary is responsible for the performance of the line managers; regional officials are responsible for the performance of the subsidiaries; and a president of the regional company in New York is responsible for the performance of the regional officers. In fact, two officers in New York will generally be responsible for regional performance. One will be designated on a geographic basis and the second on a profit center or line basis, so that, for example, if the automobile line performs poorly in Nigeria, both the regional president for Africa and the vice president in charge of automobile insurance will feel responsible for improving the results.

Considerable local autonomy for many decisions does not mean that it is difficult to sense the multinational presence in the subsidiary. Indeed, managers and employees of AIG almost without exception report the strong pressure they feel to get results that are acceptable to the home office. The decisions concerning how to get the results may be significantly in their hands, but pressure to perform is certainly external. Expatriate "Mobile Overseas Personnel," or MOPs, working in AIG subsidiaries also repeatedly emphasize the pressure they feel to produce, and the importance that this has in leading to the good results that AIG achieves internationally.*

Operations of subsidiaries are monitored closely. Each branch is responsible for monthly and quarterly reports and annual budgets or production plans that project three years ahead and are revised annually. Auditing and accounting systems are revised and directed from New York as well. If monthly or quarterly results do not meet target levels, outside involvement is likely to be prompt. A technician or administrator from a regional office or from New York is likely to be sent to the subsidiary; follow-up visits by experts who spend anywhere from a few days to many months in the country are also probable. As we shall see in our later discussion of the transfer of technology, these visits play a central role in securing first-rate performance in the subsidiaries, as well as in the transfer of knowledge about the insurance industry.

When the local affiliate does not have authority to make a decision—to accept business or to settle a claim—because its authority limit has been exceeded, the involvement of the regional and the New York offices is much more pervasive.

*MOPs generally make careers of working in overseas operations of AIG. Most AIG offices in developing countries have one or more MOPs who are assigned to that office for at least two years. Other expatriates may be assigned to a local subsidiary to accomplish a specific job in a period lasting from two weeks to two years.

Sometimes, when experienced local people send recommendations to higher authorities, these authorities may simply wire approval for a decision tentatively made on the local level. On the other hand, where expertise is lacking locally, the underwriting process may almost entirely be undertaken outside the country, with the rates set and the conditions of the new business determined elsewhere. Information in these cases will be transmitted back and forth by telex. Cases arise, too, in which a local decision needing authorization outside the country may not be approved.

The underwriting philosophy of AIG is to focus attention on those areas where results are good, to seek business aggressively in areas where profits can be made, and to turn down business where performance has in the past been poor. In other insurance companies, relationships with brokers, force of habit, or simply the search for premium income may induce a company to accept business that does not promise good results. This mode of operation is frowned upon and, indeed, is eliminated when it is discovered within AIG. As a result, conflicts may develop with local personnel. If business with a broker has in the past been renewed routinely but results have deteriorated, New York may insist that the rates be raised to such a level that the broker will seek another insurance company. This is perfectly consistent with the goals of AIG, but it may run counter to the wishes of an employee in an overseas subsidiary, just as it runs counter to the wishes of a home office employcc following similar practices.

In Kenya, renewal after renewal of workmen's compensation was being offered only at very high rates by the New York office, much to the chagrin of a local representative who would have wished to preserve his good relationship with the brokers involved. In this case intervention from New York with the local subsidiary was frequent and important to the behavior of that firm. On the other hand, in Nigeria a large construction all-risk contract was won by an MOP who sent a telex suggesting rates and terms to the New York office and received a reply two days later confirming all of his suggestions. The authority to pursue these terms with the customer did not exist in Nigeria. However, the communication with New York did not affect the decision taken; the expertise for underwriting was located in Nigeria.

In addition to the mechanisms of control and accountability we have mentioned so far, AIG also provides local companies with materials, manuals, and electronic data-processing systems that they use in their day-to-day operation. Guidelines are set in underwriting and are important in maintaining the selectivity that is central to AIG's success. Manuals also exist for settling claims. We will consider the use of data-processing equipment and procedures in more detail later when we treat the transfer of technology.

## SUMMARY

It is important to recognize that even within each of the areas outlined above, there are demands in addition to those on the technical skills of employees. Insurance technology consists not only of specialized actuarial analysis and the assessment of engineering data. These are indeed areas where competitive edges may be built. But insurance technology consists very much of the mundane, day-to-day tasks of organizing, controlling, and running efficient underwriting, claims, finance, accounting, reporting, sales, and administration departments. A company exhibits a superior technology—that is, it is able to make use of its inputs to generate more or better output than its competitors—when it undertakes these typical functions more efficiently than its competitors. Superior technology in insurance consists mainly of superior organization and operation of a large service system.

AIG has a competitive advantage in developing countries not only because of the technology it employs locally. Its advantages stem, in part, from worldwide economies of scale. It is able to use experts in various specialized lines of business to service more than one national market; it is able to use electronic data-processing equipment in one country to handle claims in another country. For example, computers in the Philippines have been used to service claims in Taiwan after a typhoon there. Advantages that derive from the scale of the operation are not readily established within a smaller organization. Similarly, AIG possesses a competitive advantage by being a worldwide company. Therefore, it may have an advantage over local companies when doing business with other multinationals or other large firms that look for a name and a reputation. To the extent that AIG's competitive advantage in developing countries rests on its worldwide spread or on economies of scale, these advantages cannot easily be transferred to insurance employees within the Third World countries. These real advantages may substantially benefit the host country, nonetheless.

However, much of AIG's technology is readily transferable. To the extent that it has highly experienced, skilled people working in close contact with citizens of less developed countries, to the extent that it has superior systems for organizations and control, to the extent to which it performs the tasks of an insurance company better, more effectively, more skillfully than its competitors, AIG is in a position to pass on the technology, the information on how this can be done, to its employees within the Third World Countries. As we shall see, the way AIG is organized in Nigeria, Kenya, and the Philippines leads to substantial transfers of technology.

## MECHANISMS OF THE TRANSFER OF TECHNOLOGY

### Formal Training

An important obvious source of technology transfer is formal training either within the firm or in sponsored institutions. AIG has a substantial in-house training program in the Philippines. It provides separate classes for top, middle, and lower-level management, and, in addition, offers correspondence classes for other personnel. These training programs are extremely intensive, involving eight hours of class per day, five or even six days a week, for a period of four weeks. Classes cover a broad variety of subjects within insurance and management. Most of the lecturers on specific areas, such as fire, marine hull, and engineering insurance, come from within AIG. They are often regional personnel based in either the Philippines or Hong Kong. In addition, experts are brought in from local universities to conduct the classes in management. In 1979, 40 employees took the management training courses and an additional 120 participated in correspondence courses. The correspondence courses are less intensive and aim principally to prepare students for qualifying examinations offered by the Insurance Institute of America. The top level management courses can be even more demanding than the middle level courses. Everyone interviewed who had participated in these courses found them to be exhausting, but also rewarding. Although participants believed they were exposed to more material than they were able to absorb fully in the time allowed, they nevertheless felt they had benefited substantially from participation in the courses run within AIG. The participants in these programs are drawn from management all over Asia. Other local firms besides AIG sometimes send employees.

In addition to the in-house training programs offered by AIG, the company finances employees' training in other institutions. Within the Philippines there is an Asian Institute of Insurance and also a Philippine Insurance Institute. In Kenya a training institution has also been opened for insurance personnel. In addition, a number of correspondence courses are offered from London, the United States, and Australia for those interested in sitting for examinations in underwriting, actuarial science, and other aspects of insurance. AIG employees are encouraged to train at these institutions; if they successfully pass an examination, AIG will reimburse them for the expense of preparing for that examination, including the cost of the correspondence courses.

AIG is also instituting training programs for agents in life insurance. These programs are less technical and less concerned with management than are the programs discussed so far. The training programs for life insurance agents are principally programs in marketing and selling. A full-year program has been developed in Wilmington, Delaware (the headquarters of the American Life Insurance Company subsidiary of AIG). The program alternates periods of training with periods of practice selling. It involves the attainment of certain targets of total business for subperiods within this year.

One of AIG's major sources of competitive advantage in all three countries visited is the organization of its selling force in life insurance. In Nigeria and Kenya, which were dominated in their colonial period by British insurance firms, the agency selling system was not well developed before AIG arrived. A large corps of field-based agents paid on a commission basis is a most effective technology for selling life insurance. AIG has gained competitive advantage, and set an important example, by introducing this U.S. marketing system.

In the Philippines, too, AIG took the initiative through the Philippine American Life Insurance Company in setting up an agency system operating throughout the Philippines, effectively managed from Manila. AIG is trying to improve the training as well as the selection program for its agents in order to maintain the competitive edge that it has built up in marketing life insurance. The training program that is being introduced into Nigeria and the Philippines should contribute to this effort.

**Contact with Experts**

It would be a mistake, however, to think of formal programs involving technical training as the principal avenue of the transfer of technology within AIG. Rather, the major way in which local employees learn AIG's technology of selling and delivering insurance is through working within AIG, in particular, through interaction with experts who work within the company. These experts may be located within the country in question; they may be located elsewhere and travel to the country occasionally; or they may be located outside the country and communicate through the post or electronic media.

Each of the three countries visited in this study made considerable use of expatriate experts. Even the Philippine American Life Insurance Company, the most self-sustaining and independent firm visited, was installing a new electronic data-processing system that involved input of outside expertise.

Contact with experts takes several forms. First, in each of these three countries, several expatriates were affiliated with each firm either in direct roles or as regional experts assigned a large proportion of the time to the firm in question. These expatriates included both general administrators and technicians. The Nigerian affiliate, both the life and the nonlife branch of the Kenya affiliate, and the American International Underwriters Philippines branch all had U.S. general managers. Nigeria also had an MOP underwriter in nonlife insurance. In Kenya the life insurance branch had another U.S. technician in charge of organizing the agency system and a British expatriate who was the second-in-command in the general insurance division. In the Philippines there were many expatriates with specific technical line responsibilities who operated on a regional level with responsibilities not only for the Philippines but also for Thailand, Indonesia, Guam, Singapore, and other countries in Southeast Asia.

In some cases expatriates have been sent over specifically to train local people. For example, the U.S. marketing director in life insurance in Kenya was instructed to find and train a replacement for himself. The regional property line director in the Philippines also was sent in to take care of a particular problem involving the quality of local underwriting and engineering, in an interesting case to which we will return shortly.

Most of the expatriates, however, are simply instructed to perform their jobs. They are rewarded when the results are good, when premiums grow rapidly, and when profits are being made. Nevertheless, in these cases too, a very substantial amount of training and transfer of technology takes place simply because the good performance of each of the expatriates depends upon the good performance of coworkers and subordinates. Thus, although a number of local people in the countries visited may not recognize that "expats" are interested in transferring skills, most acknowledged that they benefited from working together with them.

Very often within AIG, expatriate experts will be sent in to deal with a specific problem. Property insurance in the Philippines is a good example of this. Although the results in the fire insurance line in the Philippines were not bad prior to 1979, some of the local executives knew that the underwriting was not well done. In 1978 and 1979 a rash of fires (of questionable origin) drew attention to the problem in underwriting that had existed for some time. New York, reacting to the rise in the loss ratio, sent a Briton to improve the fire line's performance within the Philippines. He discovered that the underwriting problem was very substantial indeed. Not only was there a shortage of engineers, but even if engineers' reports had been well written, the underwriters were not adequately trained to read them. Therefore, several experts were sent in for five weeks each to work with local engineers to upgrade their skills. New guidelines were set for underwriting in the fire department by these expatriates. In addition, the Briton in charge of the fire line ran a course for local underwriters, which met on Saturdays, to improve their ability to handle information. Several Filipinos will also be sent to New York to work with top level underwriters within the AIG organization.

Expatriates sometimes are sent in for an extended period not in response to a crisis in underwriting or claims management, but to improve the underwriting capacity within the local market. Thus, for example, in Nigeria an expert in construction all-risk, a very technical line, spent 18 months in 1977 and 1978 training local people. Ironically, his chief trainee left AIG shortly thereafter because his skills were so marketable. Nigeria also was visited for almost two years in 1978 and 1979 by an expert in marine insurance who set up the local marine department.

In addition, each subsidiary is visited several times every year by line managers or a regional president from New York or regional offices (see Table 2.3). During these visits claims managers, underwriting managers, and other local experts go over their results and their records in detail with the visitors. Their performance is thoroughly reviewed, as well as the performance of the department for which they are responsible. Many of the line directors and other local executives in the AIG subsidiaries also had made at least one visit to New York within the last three years. These visits serve the function of familiarizing the employees with the headquarters, of providing an opportunity to discuss their operations with the people with whom they communicate on the telex, and of providing a fringe benefit of travel and an inducement for the loyalty of the employees.

These contacts with expatriates—the contact with the long-term residents, the contact with short-term visitors responsible for "fire fighting" or for setting up local department contact with the very short-term visitors who come to

**TABLE 2.3. Short-Term Visits to American Life Insurance Company (Nigeria) by Expatriate Experts (1978 and 1979)**

| | *Less than One Week* | | *More than One Week* | | *Totals* | |
|---|---|---|---|---|---|---|
| *Purpose* | *1979* | *1978* | *1979* | *1978* | *1979* | *1978* |
| Routine consultation | 5 | 7 | 0 | 1 | 5 | 8 |
| Underwriting/claims | 6 | 6 | 2 | 2 | 8 | 8 |
| Finance/accounting | 1 | 1 | 1 | 0 | 2 | 1 |
| Electronic data processing | 1 | 1 | 0 | 2 | 1 | 3 |
| Marketing | 1 | 0 | 1 | 0 | 2 | 0 |
| Administration | 1 | 2 | 0 | 2 | 1 | 4 |
| Totals | 15 | 17 | 4 | 7 | 19 | 24 |

*Source*: American Life Insurance Company records.

monitor the operation and go over records, and the contact made when local employees travel overseas—are crucially important in building an understanding among employees of the expectations within AIG for performance and procedures.

## International Communication

An extremely important mechanism for the transfer of technology is telex communication between local employees and other parts of the multinational. If employees in the subsidiary lack authorization to make final decisions themselves in certain situations, then the telex is a means of transmitting problems or recommendations for external decision making.

Whether or not the exchange of information involved in such telex communication is useful to the people in the subsidiary depends very much on their own initiative. In the general insurance branch of one subsidiary, we saw little evidence that the information involved in reporting and transmitting underwriting and claims information was useful to the local participants. They exhibited little interest in what they were doing, viewing these activities as clerical drudgery and expressing considerable frustration at their lack of deeper understanding. The expatriates in charge of this operation too expressed frustration at the lack of understanding shown by their subordinates, noting that figures had to be doubly and triply checked before they could be forwarded or used.

On the other hand, a well-trained and ambitious person making use of the telex communication system can gain a great deal from it, because the amount of information transmitted is enormous. As Table 2.4 shows, the exchange of telexes is voluminous and covers a large variety of purposes, principally underwriting and claims management. An example of how this information can be used to the advantage of the local employees is found in the aviation and casualty section of the Philippine American General Insurance Company. The aviation department, in particular, has in recent years had very limited authority to write insurance. This is due in part to a shortage of local expertise—with a chronic problem of employees skilled in aviation underwriting leaving for other companies—and in part to the concentration of aviation insurance decision making in London. Nevertheless, the person in charge of aviation and casualty at Philippine American General has shown a great deal of interest and has systematically looked at the quotes/rates supplied by AIG London. He has thus gradually increased the frequency with which he makes suggestions of rates in the telexes he sends to London. He has been able increasingly to make these suggestions because he has used the large amount of information in the telexes.

The head of AIG aviation, who is based in London, appreciated the intelligence and initiative of this employee and brought him to London for ten days in 1979 to go over files and procedures and to meet brokers in the aviation line. As a result of this employee's increase in underwriting skill, his underwriting authority in the Philippines has been raised, so fewer contacts with AIG London are required today in aviation underwriting. It should be noted, however, that underwriting authority still is rather limited because so much of the decision making is concentrated in the London market. Nevertheless, the increase in this employee's skills and the consequent reduction in delays in processing requests for aviation

**TABLE 2.4. Telex Records: Nigeria and Kenya**

| | *Nigeria*[a] | *Kenya*[b] | |
|---|---|---|---|
| | *Week of 18 Feb* | *Week of 3 March 1980* | |
| *Subject* | *1980* | *Inward* | *Outward* |
| Underwriting | 7 | 13 | 10 |
| Claims | 2 | 2 | 3 |
| Accounting | 2 | 2 | 1 |
| Finance | 3 | — | — |
| Administration | 1 | — | — |
| Electronic data processing | 1 | — | — |
| Destination[c] | | | |
| New York | 9 | 9 | 7 |
| London | 2 | 7 | 5 |
| Other | 5 | 1 | 1 |

[a]Outgoing only.

[b]General insurance only.

[c]All but two of the telexes in Nigeria and four in Kenya were internal AIG communications.

*Source*: AIG, Company records.

insurance have benefited AIG, which receives more business as a result. This is an example of a typical transfer of technology beneficial both to the employee and the company and illustrates a role often played by international telex communication.

Not all telex communication is mandated by the company because of limited authorization in the subsidiary. The head of marine insurance at one of AIG's Philippine firms provides an example of the voluntary use of telex for learning. In his early days as the chief marine underwriter in the company, this employee was very nervous about making decisions, afraid he would make a mistake and lose the authority he had been granted and perhaps even his job. He ironically refers to himself today as having been a "telex clerk" then. Afraid of making mistakes, he asked advice on almost every decision he had to make, including routine decisions, even though this was not required of him. In time, however, he was able to anticipate replies to routine telexes so well that he stopped sending them. Now, more than a decade later, his operation is quite autonomous, with all but a few very large or complicated policies being handled by him without consultation with the rest of the company.

**Systems and Procedures**

Much of the transfer of technology in insurance involves the increase in the skills of employees through their experience working in and running a high-powered insurance operation. Another significant form of transfer of technology, however, comes through the direct transfer of reporting systems, procedures, products, and, particularly, electronic data-processing equipment and systems from the parent company to its subsidiaries. We already have discussed how the Philippine American Life Insurance Company adopted the Japanese procedure in claims adjudication and how this has earned them considerable good will at low cost. Other similar examples abound. Most of these do not involve sophisticated technology or products, but nevertheless they show how access to a large and changing set of ideas and products improves production and output in local subsidiaries.

In Nigeria, for example, two years ago a life insurance policy was selected from the AIG portfolio by local employees because it seemed to meet the needs and tastes of Nigerians better than policies that had been sold earlier. This "modified anticipated endowment" policy long had been available among American Life Insurance Company policies, but it had not been pushed by expatriate employees. Local employees recognized that Nigerians prefer a policy that offers them something during their lifetime as well as something to their beneficiaries after their death. These employees, therefore, decided to promote this policy and have seen it grow rapidly to one of their most popular products. This is an example of a two-way transfer of ideas. The details of the policy were transmitted from the American Life Insurance Company to the Nigerian operation. However, the decision to choose this policy was made by Nigerian employees who had a better feel for the market and were able to use this to the benefit of the company.

The best example of sophisticated technology transmitted from other branches of AIG to Third World companies is to be found in electronic data-processing systems. AIG is pushing hard on this front, adopting new electronic data-processing equipment and systems both in the field and at its home office. Some of these systems, particularly a new operating system with which subsidiaries will be called upon to supply much more information than in the past to the home office, were not popular in the countries studied. This Overseas General Insurance System was viewed as a move toward centralization of decision making and a demand upon resources without any obvious payoff.

Most electronic data-processing systems, however, are very much appreciated, and even those persons within AIG with strong nationalistic feeling recognize the importance of the AIG contribution to local computer capabilities. For example, in the Philippine American Life Insurance Company, which for the most part is an autonomous, well-run, locally managed company, the electronic data-processing system is being substantially revised and expanded with a large input of external resources. A Dallas company is installing a new system at a cost of more than $1.25 million. It is also sending a representative for a year and a half to provide system support. AIG is sending three Filipinos to Dallas, New York, and Wilmington for several months each for training in the use of the new computer facility. A similar major investment was made in Nigeria in 1978 (before the Nigerian government takeover).

Although the systems are being purchased from outside agents, the multinational connection through AIG is important. First of all, it is relatively easy for the multinational to make the substantial investment required to purchase the machinery and to train people in its operation. This investment would be much more difficult for a smaller local company. Second, the multinational, through setting up systems simultaneously in several countries, is able to get quantity discounts from the suppliers of these facilities—in this case, the Dallas-based computing company. Finally, because it is a large operation with a great deal of potential business in the future and many employees all over the world, AIG is able to put pressure on its suppliers to meet contracts and to perform as specified in agreements. They make the equipment more easily available in the Third World countries; they also ensure that people will be trained to use the equipment properly, and that the equipment will produce a good return on the initial investment.

Some of the systems and procedures used by AIG may contribute to the transfer of technology not only in being in themselves a useful technology, but also in helping employees learn more about insurance. Particularly important in this regard is the reporting system within the corporation. Intelligent and experienced employees in the field when filling out a report form will anticipate questions from the home office. They are likely to prepare themselves for these questions by investigating beforehand the causes of poor performance or unexpected results. Thus, the format of reporting gets employees to ask the right questions, just as consultations later help them find the right answers. One young and capable Nigerian underwriter specifically pointed to the reporting procedure as a valuable system in his development as an underwriter. He said that filling out the forms helped him to see what was going on and to find problems before they were pointed out to him by his supervisors.

On the other hand, if the background and experience of underwriters are inadequate or if their general education does not prepare them to handle this type of question, the reporting procedures may be of no use to them. The underwriting manager in one firm, for example, looked upon reporting as just one more bureaucratic chore and was resentful of the questions that followed the results reported by him. He did not know the answers to the questions and appeared not be stimulated to seek them by the procedure of filling out reports.

The contrast between the role the same report plays in the transfer of technology in these two examples emphasizes the importance of employing well-trained and skilled people to begin with. If people lack the general education and the background to make use of the systems, procedures, and materials available to them, then not only will there be a minimal transfer of technology, but these people's performance is likely to be weak as well. It is interesting to note, as is evident in Table 2.5, that the general level of education in the Kenya general insurance operation was significantly lower than that in the life insurance branch or in the Nigerian insurance company.

**TABLE 2.5. Education of African Head Office Employees in Nigeria and Kenya[a]**

| | *Nigeria* | *Kenya (%)* | |
|---|---|---|---|
| | *(%)* | *Life[b]* | *General* |
| University | 10 | 11 | 5 |
| "A" levels (advanced high school) | 2 | 19 | 9 |
| "O" levels (ordinary high school) | 74 | 57 | 72 |
| Other | 14 | 13 | 14 |

[a]In the Philippines, almost all employees of AIG have a university education, although the quality of universities varies greatly.

[b]The life insurance operation in Kenya now hires only people with at least two "A" level passes.

*Source*: Questionnaire responses.

Some Kenyan executives appeared to have difficulties in performing to standard, which seemed traceable to their inadequate general educational background and their consequent inability to make use of the materials at their disposal.

Through pursuing its own profits, then, AIG finds it in its own interest to train local people, to expect of them high level performance, to monitor their work and correct it when it is in error, to provide them with materials and systems that lead to high productivity, and to procure sophisticated electronic data-processing equipment and train people to use it. All of these constitute the transfer of technology. Very little of this transfer is undertaken specifically with an eye toward transferring technology, yet this is the general result one observes.

## Constraints

These mechanisms are vital to the transfer of insurance technology, but they are not problemfree. It is certainly true that there are cases in which the parent company has attempted to introduce systems or approaches that do not fit well within the local cultural setting. Any company or individual involved in transferring technology has the difficult job of constantly deciding where corporate goals and approaches will work and where they may have to be modified to achieve success in the local setting. Several examples from the countries studied illustrate the ways in which systems have been adapted when problems were encountered as well as the difficulty of deciding when such adaptation should be made.

In the Philippines, the practice of demanding competitive bids on potential auto repairs does not work effectively. Eventually, a new system in the automobile repair line was introduced through the employment of in-house appraisers who work in agreement with certain shops. This new organization appears to be a better way to get the job done under the local circumstances.

Nigeria offers another example of an inappropriate system, this in the marketing of life insurance. In 1979 a new marketing system was introduced from New York that had worked well in the United States and the Caribbean. Several seminars were held, but the system met with almost no success. The reason for this failure was the inappropriateness of the marketing approach for Nigerians. The emphasis of the marketing plan was on nuclear families, on retirement, on a suburban life-style. Nigerians did not find the "message" attractive and have not pursued it. The attempt to introduce this sales approach was, in retrospect, costly and ill suited to local conditions.

Sometimes, however, a way of doing business that seems inappropriate for local conditions may be absolutely essential and greatly improve the quality of performance. For example, a young Filipino stated that one of the most important lessons he learned while working with expatriates within the organization was how to say "no" when an application is made for a policy. The U.S. style tends to be blunt and direct compared to

Filipino style. This can lead to personality clashes that undermine an effective organization. On the other hand, it can lead to better underwriting if people are told directly that they will not be insured. One of the great problems of AIG's general insurance branch in Kenya is that some Kenyans find it difficult to turn down brokers or to put pressure on people with whom they have been dealing in the past and with whom they expect to deal in the future. In this case, an important part of underwriting, selecting between good and bad risks, is undermined by traditions of behavior in the local business community. In the Philippines a similar problem exists. Whether or not local people like the bluntness of U.S. or other expatriates, it sometime gets the job done.

It is not always easy to determine when a technology is appropriate. The home office may insist that a procedure be followed, much to the dislike of the employees in the subsidiaries, to find in the long run that the mandated procedure has worked well and has accomplished what it was intended to accomplish. On the other hand, the mandated procedure may turn out simply not to apply to local conditions and the effort put into trying to impose it may turn out to have been wasted. With management styles, too—an important part of the technology of running service organizations—it may turn out that an unpopular style simply reduces morale and effectiveness of the work force, or it may turn out that this style is very effective in generating results. Recognizing the difference requires experience, sensitivity, and a fair share of good luck.

## TRANSFER OF TECHNOLOGY BEYOND THE SUBSIDIARY

The transfer of technology does not stop at the firm level. One of the most frequent ways in which AIG transfers technology beyond its own subsidiaries is through the loss of skilled people to other employers. Because the insurance industry is skill-intensive and because skills tend to be very scarce in developing economies, a first-rate insurance seller is likely to have many opportunities.

Although it was difficult to secure accurate statistics, it was possible to obtain a substantial list both of AIG employees who had left the companies visited and of current employees who had been hired away from other companies. In Kenya, AIG does not appear to be a net "exporter" of trained people to other firms. Nobody has left the life insurance branch in the last few years, and only one skilled employee has left the nonlife branch. Thus, most of the upgrading of skills and improvement of the quality of the human capital has benefited AIG in Kenya. In Nigeria, and especially in the Philippines, where AIG is an older, more established, and more important firm in the local market, AIG appears to be a very substantial net exporter of trained people. To give two examples: Within the last few years, no fewer than ten prospective successors to the vice president in charge of property and fire in the AIG branch in the Philippines have been hired away by other companies. The Philippine American General Insurance Company, too, lost four heads of aviation underwriting between 1970 and 1980, all to other companies. In the Philippines, AIG is known as the "training ground" for the insurance industry.

However, there is also a small flow in the other direction, into AIG from other companies. This flow can benefit AIG. For example, in Kenya, when the chief accountant of the general insurance branch was hired from another company, he introduced a system learned in his previous job for reconciling the company accounts with those of brokers.

Although such a flow of trained people within the industry helps spread good insurance practices and upgrades the technology within the industry as a whole, the outflow of skilled people can at times be a hindrance to the spread of technology into the subsidiary itself. AIG indicated that it was particularly interested in training its Japanese employees because of the implicit long-term contract that exists in labor markets in Japan. AIG felt confident that it would get a good return on the investment made in training people locally and in New York. Equal confidence could not be held in the Philippines or in Nigeria, where people are likely to take their skills, often acquired at company expense, into the market to earn a higher return.

In addition to exporting skilled people, AIG has provided other benefits to the insurance industry by setting examples and by collaborating with insurance companies in coinsurance and in other cooperative ventures. For example, when the Nigeria National Insurance Company, a state-owned company, was developing its claims department in 1979, it sent people to interview the Nigerian claims manager at AIG; several afternoons were spent working carefully on issues concerning the establishment of an effective claims department. In Kenya, AIG worked with the Kenya National Insurance Company on oil pipeline coverage. AIG provided the basic underwriting skills, although the Kenya company was in charge of this undertaking. In these cases the skills provided by AIG probably would also have been available from reinsurance companies or other consultative sources, but the fact that AIG is so easily accessible made the collaboration attractive for the national insurance companies and provided readily available skills and technology to the local economy.

## THE MULTINATIONAL CORPORATION AS A SOURCE OF TECHNOLOGY

MNCs are only one source of technology. If insurance technology is divided into skills and procedures, on the one hand, and management and organization, on the other, then the former certainly can be garnered from a large variety of training institutions and schools. Underwriting, actuarial science, and financial management are all subjects that can be learned in colleges of insurance and business. Indeed, there is hardly a developing country that does not have a training school for insurance personnel. Similarly, the United Nations Conference on Trade and Development is supporting training institutions all over the world. Local citizens also can be sent overseas to study or to work at an insurance company and learn skills and tasks on the job. Technical services can be imported as well. A number of large reinsurance companies, Swiss Reinsurance, in particular, offer skilled engineering underwriting services to their clients. Thus, some aspects of technology certainly can be learned directly on an open market without giving a multinational firm access to the local direct insurance market. To some extent, management and organization systems, too, can be acquired through the study of the organization of other firms or through training in business schools.

The advantages of the MNC operating within the country as a source of technology are great, however, and this explains in large part the continued presence of multinationals, even in environments in which they are resented. As we have seen, a large proportion of the total transfer of skills and knowledge within the insurance business takes place not in formal training sessions nor in easily isolated blocks of time, but rather in continuous, long-term interaction of people within a well-organized framework of accountability, with high standards for technical performance. After a number of years, an institution can be built in which the workers and the managers are performing smoothly, carrying out functions in the way expected by the home office.

For example, Philippine American Life Insurance Company is the largest life insurance company in the Philippines. It is headed by two Filipinos who have more than 25 years of experience with the company. Virtually all major positions are held by Filipinos. Although the company is not self-sufficient, particularly with regard to data processing and computer use, local people know how to run the company at the managerial and technical levels. Their primary training has come through their continuous relationship with their parent company, AIG. They have been trained by universities. Some have received substantial training while working for other insurance firms, but almost all owe much of their expertise to learning within the firm. Although AIG did not go to the Philippines 30 years ago to transfer technology, the conclusion is inescapable that this has indeed happened and continues to happen.

AIG has excellent technicians, management systems, and reporting systems, which together make it an efficient and profitable organization. The advantage to the host country in having a company such as AIG operating within its borders is twofold: It sets an example, and it trains local people in the process of performing the job efficiently. A smoothly functioning multinational shows the "state of the art." For a number of technical procedures, this state of the art can be learned through academic training. It may be learned from consultants for the organization of a particular department or division. But for an entire firm, it can best, and perhaps only, be learned by working within the system. Thus, an MNC has an almost indispensable advantage for transferring technology in a service industry, for it can draw on global experience to institute and monitor organization and procedures.

MNCs have an additional advantage in economies of scale. Their worldwide scope makes it possible to develop and install advanced computer systems and to train and deploy highly specialized experts. No individual firm, serving a limited market, could afford to develop such services. Although they can be contracted for individually, they would not then be integrated into the local company system. It is this integration of skills and techniques into an ongoing system that leads to profitability and efficiency.

Much of the transfer of technology within AIG to Third World countries could easily be overlooked. Only a relatively small part involves big computers, oil rig underwriting, new methods of actuarial analysis, and the like. Yet the day-to-day control and management of the MNCs lead to steady improvement in skill levels in the local country. The process of building a local insurance company is not simple, nor is it easy to accelerate. Disagreements and differences in approach between the multinational and local levels can be expected. Nevertheless, in the long run, with effective policy at the local level and foresight on the part of the multinational, it can work out fruitfully for all parties concerned.

CHAPTER 3

# Mass Merchandising and Economic Development: Sears, Roebuck and Co. in Mexico and Peru

*Nancy Sherwood Truitt*

*"Marketing is . . . the process through which economy is integrated into society to serve human needs."*
*—Peter F. Drucker**

## SEARS, ROEBUCK AND CO. IN MEXICO AND PERU[1]

### Introduction

Until recently, little attention has been paid in economic development theory to the role of the service sector. Concentration has been placed on industrialization, virtually to the exclusion of both agriculture and services. As a result, the role of the service sector in economic and social development has been widely overlooked. This chapter will focus on one particular component of the service sector, that of commerce, and will argue that the marketing function and the technologies necessary to support it are a vital component in the process of economic and social development.

In a traditional economy, the commercial sector serves simply as an intermediary between the producer and the

*From "Marketing and Economic Development," *The Journal of Marketing,* January 1958.

consumer. The concept of the market is that of a static rather than dynamic one. It is this concept of the market that has significant ramifications in terms of economic development.

Whereas capital is generally viewed as a limiting factor in achieving development, the technology and managerial skills necessary to use existing resources efficiently are of equal importance. The introduction of the modern marketing concept of mass merchandising into Mexico and Peru by Sears, Roebuck and Co. provides an excellent example. Capital was not the major limiting factor in expanding the market; locally owned department stores existed in both countries, serving an elite trade. Missing was the concept of mass merchandising and the technical and managerial skills necessary to implement it.

It has generally been assumed that changes in market structure and marketing policies come as a result of industrialization, not that they are initiating sources of growth and productivity.[2] The thesis of this study is that the introduction of the concept of marketing has helped change static markets into dynamic ones, and that this change has initiated an upward spiral of employment and purchasing power while at the same time making more efficient use of existing resources.

## Mass Merchandising

Mass merchandising is a development of the twentieth century. It represents the application of modern management technologies to the commercial sector. It has revolutionized that sector as other technologies have revolutionized the industrial and agricultural sectors. Mass merchandising provides the key link between consumers and producers that enables the producers to know what the consumers want and how much they will buy at what price levels. Without this link, resources in the form of capital, labor, and entrepreneurial ability may be applied to producing items that have little market; or, as more often happens, large market opportunities lie unexploited because no one probes their existence. In either case, the result is inadequate use of resources and loss of economic growth potential.

Several crucial components of mass merchandising distinguish it from traditional retailing. Mass merchandising has been built on the concept of making satisfactory profits through low margins and high volumes of sales rather than high markups at low volumes. Therefore, it has played a significant role in bringing quality goods to previously excluded segments of society. As its name indicates, it is merchandising for the masses: the systematic location and development of supply sources of products desired by consumers and made available in the quantity and quality desired, at an affordable price and with a guarantee of product performance and service. In contrast, traditional retailing has been characterized by a wide variety of specialty shops with limited selection of products. Department stores, where they exist, are primarily for the elite trade. As a result, economies of scale are generally not available. The wholesale cost of the product, overhead, transport, and other factors tends to be high per product sold.

A second major distinction lies in the fact that mass merchandising links the consumer backward to the sources of supply. A traditional distribution system must rely for its merchandise on the product that manufacturers have decided to produce. Retailers have limited control over product type or quality, since they can select only from among the available supply. The mass merchandiser, on the other hand, engages in research to determine what the consumer wants and then translates those wants into products by working directly with the manufacturers. This relationship with manufacturers, called product and source development, may include, in addition to product ideas and improvements, technical and managerial assistance, quality control, and even financing. By maintaining a direct relationship with both the manufacturer and the consumer, the mass merchandiser creates a direct link between the two. In contrast, the national retailer relates only to the consumer, having no role in product development.

The means of building and maintaining consumer confidence is the third distinction between mass merchandising and traditional retailing. Because it is oriented toward a mass market, mass merchandising cannot depend on developing the personal relations with each customer that characterize the many in the traditional retail sector. Yet consumer confidence is crucial to success. To build and maintain consumer confidence, the mass merchandiser depends on quality and service at reasonable prices, product guarantees, service and repair facilities, product information, and well-trained, courteous personnel.

Mass merchandising was basically developed by Sears, Roebuck and Co. in the United States. As Peter Drucker has pointed out in his cogent analysis of Sears, "The Sears policy, all along, has been to find the majority market and to convert it into a true mass market."[3] It took innovations in traditional retailing practice to do this. Catalogs were the means of access to the dispersed and untapped rural markets of the early 1900s—an initiative repeated in later decades by locating stores (with parking lots) in the outskirts of the cities to serve "the motorized farmer and the urban population."[4]

Equally important as site location in developing a mass market is the alliance of consumer research, product design, and manufacturer development, which Sears pioneered in order to provide a selection of products at affordable prices. Another innovation was the Sears policy of "satisfaction guaranteed or your money back." Crucial to implementing innovations such as these successfully is a well-functioning organization; again, an area in which Sears pioneered, with management development programs and profit sharing.[5]

This chapter will describe the operations of Sears, Roebuck and Co. in Mexico and Peru as a basis for drawing conclusions about the impact of mass merchandising on economic development. It will first review the history of Sears entry into these countries, then describe the technologies of mass merchandising that Sears introduced, and finally discuss the role of the marketing concept of mass merchandising in economic development.

## Introducing Mass Merchandising in Latin America

Sears went international because of the "vision of General Wood,"[6] the long-time leader of Sears who moved it from a catalog operation to a retailing giant. General Wood had particular interest in Mexico and Central America. He believed that Latin America represented an area of real economic potential, and pushed the expansion of Sears first into Cuba in 1942 (later nationalized by Castro) and then into Mexico after World War II. This was followed over the years by the establishment of stores in a total of 11 Latin American countries from Brazil to Mexico. By 1978 Sears had 128 retail outlets in Latin America and Spain, employing over 23,000 men and women and administered through 11 subsidiaries. (The only Sears affiliate in Europe is its moderately successful subsidiary in Spain, after an unsuccessful joint venture in Belgium. It has a successful joint venture, Simpsons-Sears, in Canada.)

However, as a recent chronicler of Sears history pointed out, Sears is a "giant: that would never become a multinational. . . its future would be wholly determined by how well it performed in the American market. . . ."[7] An analysis of the statistics bears this out. In 1978 net sales from Latin America represented 3 percent of the total net sales of Sears and 2.5 percent of net income. Thus, Sears is very much a domestically oriented company.

The relatively small size of its international operations means that only a limited commitment, in overall corporate terms, can be made to adapting the corporation to fit foreign environments. What works in the United States, with some modifications, is expected to work abroad, at least in Latin America. The fascinating thing is that this strategy has succeeded to date, as this and other studies document. The question for the future is whether modifications and adaptations of Sears U.S. will continue to meet the needs of Latin markets.

In terms of size, population, stage of development, economic conditions, and political structure, Mexico and Peru present several interesting contrasts. Yet, despite their differences, Sears has had a similar impact in both in terms of modernizing the commercial sector and promoting local industry.

A few comparisons set the scene. In terms of population, gross domestic product, and market size, Mexico represented some 18 to 22 percent of the Latin American totals in 1977, whereas Peru accounted for only 3.5 percent.[8] Per capita income in Mexico was $977, whereas that in Peru was $723.

But perhaps most telling for a study examining the development of local markets is information on purchasing power and market growth. An estimate of the "richness," or degree of concentrated purchasing power, of a market as compared with the Latin American regional average of 1.0 puts Mexico at 1.04, slightly above average, whereas Peru's has been 0.31.[9] Thus, Mexico represents a much larger and richer market than Peru.

Further, Mexico has had a stable government since the early 1920s. The economic policy of successive administrations has been to promote rapid industrialization primarily through import substitution policies. With its large population, good transportation, and numerous well-developed cities, a broad industrial base has been developed. It has experienced an excellent rate of growth over the three decades from 1940 to 1970, although growth has slowed since then. Great hopes for continued expansion and enhanced ability to deal with the significant poverty that still exists now reside in putting the income from Mexico's extensive oil and gas deposits to effective use. However, income continues to be unevenly distributed.[10]

Along with industrialization, the government has been pursuing a policy of Mexicanization of industry, which was codified in the Foreign Investment Law of 1973. Among its provisions is a limitation of foreign ownership to 49 percent for any new company established in Mexico and the establishment of a national commission to authorize and control all foreign investment. How this law reducing new foreign ownership to a minority position would be applied to expansion of existing operations was one of the key factors holding up Sears expansion in Mexico in the mid-1970s.

Peru has been characterized by political instability, with a history of military intervention. Its economy has always been heavily export oriented, and the country demonstrates the enclave development typical of such an economy. Wealth is concentrated in the upper and middle class areas of Lima; the vast majority of the country lives in rural or urban poverty, many of them outside the monetary economy.

In an attempt to reduce foreign dependence and to industrialize, Peru adopted an import substitution policy in the early 1960s. This approach continued to be followed by the military who overthrew the civilian government in 1968. The military took power determined to develop the nation's industrial base while better distributing the fruits of economic activity, since Peru had one of the most regressive income distributions in Latin America. The government set out to revolutionize the economic structure of the country, taking control of basic industries, promoting worker ownership of half of each privately held industrial enterprise, encouraging a cooperative industrial sector, and redistributing ownership of agricultural lands.

In dealing with foreign investment, Peru strictly applied the Andean Common Market policy of limiting the role of foreign capital. These measures, combined with the implementation of the revolutionary economic and social policies, a drastic fall in export earnings, problems resulting from import substitution policies, and heavy external borrowing, led to a severe economic crisis. The rate of growth of gross domestic product in both 1977 and 1978 was negative, and Peru was close to bankruptcy. By 1978 inflation was some 80 percent per year.

In 1975 a change in government leadership took place, and a commitment was made to elections in 1980. A number of policies were revised in order to promote private investment, including reduction of workers' potential ownership of industrial enterprises to some 30 percent. At the same time, Andean Common Market regulations regarding foreign investment were liberalized. Given the severity of the present economic crisis and the upcoming elections, the future of Peru, unlike that of Mexico, continues to be unclear.

### *Sears Enters Mexico and Peru*

The introduction of mass merchandising techniques into Latin America by Sears, Roebuck and Co. had similar effects to those seen in the United States. The traditional retailing sectors were forced by the competition to adopt similar techniques or lose markets. Latent demand was turned into purchasing power through making credit available to the middle and upper-middle classes. Local production of mass goods for sale meant more jobs, which were translated into increased demand. Efficiencies were introduced into the supply-and-demand equation through market research, product development, and modern marketing techniques. Essentially, the march toward a mass market for consumer goods was begun through attention to the needs of such a market.

Sears established its first store in Mexico City in 1947. Within six years the number of stores had increased to seven, six of them located outside the Federal District that encompasses Mexico City. Sears rapidly became one of the largest department stores in Mexico by introducing mass merchandising concepts. Their primary competitors were described as "catering to the carriage trade." In them, goods were not openly displayed, and prices were hidden. Quality was not guaranteed, and variety was limited.[11] The average Mexican did not shop with them but at various specialty shops and street markets. The situation was similar in Peru when Sears established its first store there in 1955. The only existing department store, locally owned, served a rather exclusive, upper class clientele.

The innovations that Sears introduced into Mexico and Peru were those that had developed in the United States: new ideas about store location, consumer research, product development, credit, and an active approach to marketing its products. Merchandise displayed in windows that could be observed from the street and lighted at night was a new concept, as was large space advertising in the newspapers. Both were rapidly copied by the competition. Where retailers had previously carefully placed merchandise under glass counters, Sears put it on open display and provided descriptions on the tags. Traditional Latin bargaining over price was replaced at Sears by set prices. The marketing concept of price lining—providing merchandise at a series of different prices representing various levels of quality or style—was introduced. The Sears promise of "satisfaction guaranteed or your money back" accompanied all products. And a radical innovation in store location was undertaken when Sears located its first store in Peru in the suburbs outside the center city.

These were marketing innovations designed to attract a larger and less exclusive clientele than that of the traditional department stores in both countries. To make it successful, two additional innovations were necessary: the provision of credit to facilitate purchases, particularly of consumer durables, and the development of dependable sources of product supply. Within five years of the establishment of the first store in Mexico City, credit sales represented 47 percent of all Sears Mexico sales.[12] A random sample study of Sears credit customers done by the National Planning Association for a 1952 study concluded that Sears customers, credit and noncredit alike, "represented a fair cross section of at least the upper 50 percent of economic groups in Mexico City."[13] Sears was beginning to reach into the mass market, one that had been untouched by the existing department stores.

Because of the differences in developmental stages, the situation in Peru was not the same. With only some 6 percent of the economically active population earning over $58 per month,[14] it would clearly have been impossible for Sears to tap as broad a market as it had in Mexico. Yet Sears believed that the market potential was there even if initially the primary market was among the upper and middle classes.[15] A major way of broadening the market is to offer credit. Within five years of its establishment, over 40 percent of its sales were made on credit, and its average credit customer was described as coming from Lima's middle class.[16] Although other stores offered credit before Sears introduced its plan, "the success of Sears credit system was impressive enough for two other retail organizations to copy most of its features."[17]

Developing dependable sources of product supply into the quality and quantity needed probably provided Sears with the most difficult problems during its first years in Mexico and Peru, and led to the introduction of one of its most important innovations, product and source development. Sears entered Mexico with the intention of relying on imported goods for a significant portion of its merchandise. Clearly, had Sears been able to follow this policy, its impact on economic development would not have been as great, because many of the jobs and skills required to produce the products would have remained outside the country.

Within a year of the opening of the first Sears store, embargoes followed by import duties were imposed by the Mexican government on a wide range of consumer goods, and Sears was forced to move massively into local product development or face probable failure. This involved everything from designing products that met Sears specifications and could be manufactured locally with available raw materials to helping establish local manufacturers. When Sears opened its first store in Mexico City, 50 percent of the products sold were purchased locally. In 1955 the figure was 70 percent, and by 1978 it was 99.8 percent. In Peru the process was similar, with 48 percent of the products being purchased locally when the first store was opened in 1955. Local purchase by 1978 was 97.9 percent, with an additional 1.16 percent coming from the Andean Common Market.

Two factors favored the broad distribution of the results of Sears product and source development efforts. One was the need for enough production volume to bring the product price to affordable levels. This generally called for more production than Sears alone could sell. The second factor was Sears preference that manufacturers not be dependent solely on Sears. As a result, many of the products Sears developed or improved came to be sold by its competitors albeit without various Sears, Roebuck's exclusive features.

Sears challenged traditional ways of retailing when it entered Mexico and Peru. To make a profit based on fast turnover of merchandise rather than high margins required introducing techniques that would result in large volume sales. And this, in turn, meant aiming at a broad market, designing products that would attract it, providing credit so goods could be bought, and promoting their purchase. Most of the techniques introduced by Sears to do this were soon copied by its competition, resulting in modernization of the retail sector.

*Sears Mexico in 1978*

Sears Mexico in 1978 had 41 stores located throughout Mexico. Four of them were large, full-line department stores known as A stores. All were located in the Federal District consisting primarily of Mexico City. Two of them, Universidad and Satelite, were anchor stores in shopping centers that Sears was instrumental in developing. There were six B stores, which are smaller in size than A stores but carry a full line of merchandise. As one would expect, only one was located in Mexico City; the rest were in major secondary cities of Mexico—Monterrey, Guadalajara, Tampico, Puebla, and Mexicali. There were 17 C stores, which generally carry only hard lines (appliances and furniture). Over 85 percent of them were located outside the Federal District, as were most of the 14 satellite stores, which carry a smaller selection of hard lines than do C stores.

Sears major growth took place in two spurts, during its first decade in Mexico, from 1947 to 1957, when two A, four B, and seven C stores and one satellite were constructed, and in the five years between 1968 and 1973, when two A, two C, and eight satellite stores were built.

In 1978 Sears faced stiff competition in Mexico, particularly in Mexico City, from several different types of stores. There were three major department store chains: El Palacio de Hierro, El Puerto de Liverpool, and Salinas y Rocha. The first two served a market generally judged to be among the upper-middle and upper classes. They were stronger in soft lines (clothing, accessories, housewares) than Sears and did not place as much emphasis on hard lines, particularly appliances, as did Sears. Neither had repair after sale service as part of their operations, but repair was available from the producer of the product. All of these department stores offered credit.

Visits to these stores confirmed that they were very effective merchandisers, apparently the equivalent of any in New York or Chicago. Liverpool had several stores in large cities outside Mexico City. Salinas y Rocha was smaller and less sophisticated than Palacio or Liverpool. It did, however, have several stores selling only hard lines in cities outside Mexico City.

Another growing type of competition on the lower end of the market was provided by the discount stores, such as Comercial Mexicana, De Todo, and Gigante, which grew rapidly in the 1960s using the discounting, self-service approach to merchandising popular in the United States, particularly with the price-conscious middle and lower classes. These stores sometimes combined soft lines with supermarkets. Most had credit available through bank credit cards. None had repair or after sale service. Comercial Mexicana had expanded outside Mexico City into a number of smaller cities.

A third approach was represented by the chain Aurrera, which started with discount stores and in 1978 was rapidly moving "up the line" into the middle and upper-middle class market. It had established furniture and appliance stores, clothing stores, and supermarkets throughout the Federal District. In a sophisticated marketing strategy, a different name was used for each type of store, serving to distinguish it from the original discount store.

Finally, there were the specialty shops selling appliances, furniture, clothing, and so forth. In general, unless they were representatives of the manufacturer, they did not have repair or after sale service. The credit available through them was primarily on the basis of bank credit cards.

In a 1976 study of the Mexican economy that listed the capital of the major retail establishments, Liverpool and Aurrera were the largest, followed by Sears, Roebuck and Co., Comercial Mexicana, Gigante, and Salinas y Rocha. Sears, of course, was foreign owned. Aurrera had 40 percent foreign ownership through the Jewel Tea Company of the United States. Other foreign involvement was via the technical assistance that several stores, such as Puerto de Liverpool, received for particular projects.

Although a great deal of information was not available, some conclusions regarding the customers of the various stores could be drawn from a 1969 market information study of retailing in Mexico City. The study indicates that Sears and Comercial Mexicana ranked as among the most important retailers to the middle class. This study identified over half of Sears shoppers as earning less than 5,000 pesos per month ($400), whereas only 35 percent of the respondents who shopped primarily at Liverpool and Palacio de Hierro earned less than this amount. As a point of comparison, the 1970 census listed 87 percent of the population in Mexico City as earning less than 5,000 pesos per month.

Whereas the first burst of expansion for Sears took place in the late 1940s and the 1950s, most of its competitors grew rapidly in the 1960s. As Douglas Lamont in his 1972 article on retailing in Mexico has pointed out:

> It [Sears] has pushed the two traditional, French-style department stores, El Palacio de Hierro and El Puerto de Liverpool, into adopting most of its merchandising innovations. The irony of it all is that in recent years, they have earned higher profits on sales than Sears, and thus they have been beating Sears at its own game.[18]

The loss of position in the Mexican market by Sears can be attributed to a corporate management oriented primarily toward the stable, predictable U.S. market and environment. Concern for the future of its investment in Mexico began with a large loss incurred in the 1954 devaluation of the peso and continued during the 1950s fueled by the anti-U.S. sentiments perceived in Mexico and elsewhere in Latin America. While Sears held back, its competitors forged ahead, using many of the techniques associated with Sears to increase their market share. Thus, by the early 1960s, when management changes brought a renewed commitment to growth in Mexico and throughout Latin America, Sears was facing aggressive, effective competition in the Mexico City market.

There is a long lead time in a new store development. Thus, the decisions of the early 1960s were not reflected in new stores until the mid- to late-1960s when 11 C and satellite stores were built and the new Universidad shopping center was opened. Other indications of efforts to increase market penetration could be found in the particular attention given to new clothing and shoe lines over the last ten years. These, of course, were the areas in which Sears faced strong competition. Special attention was also given, with Sears International assistance, to upgrading the training of credit personnel in order to improve performance and attract more customers. At the same time, Sears continued to expand outside Mexico City, and by 1977 over 50 percent of its sales came from the zone, as the area outside the Federal District of Mexico City is designated by Sears.

It appeared that Sears Mexico in 1978 was a rather stodgy, middle class store whose direction in an increasingly competive and rapidly changing market was in flux. To adopt a full-scale strategy for taking on the sophisticated markets being served by Liverpool and Palacio de Hierro at a time when foreign ownership was under increasing pressure would be unwise. Sears began to look for those focal points of opportunity in the economic development needs of Mexico where its know-how could be profitably applied. The cities and towns outside Mexico City were viewed by the president of Sears Mexico as the "major growth area" of the next 20 years. They were also the focus of the Mexican government's National Urban Development Plan, one of whose major aims it was to organize small- and medium-sized towns so as to give the local population access to the benefits of urban activities and services. This was a goal to which Sears could contribute.

That Sears would be able to respond to this opportunity seemed clear, with formal agreement having been reached between Sears and the government of Mexico regarding expansion plans. Such plans had lain in abeyance for several years until the implications of Mexico's law limiting foreign ownership of new investment to a minority position were clarified as they would apply to an expansion.

In 1978 the government of Mexico and Sears agreed to a formula that would increase Mexican ownership of Sears gradually during the coming years as Sears Mexico expanded its operations in the country. Over 80 percent of this expansion was planned to take place in less-developed towns where commerce was not well organized. Given the general attitude of the government toward foreign investment in the retail sector at that time and its willingness to approve this expansion, one can only conclude that it was looking to Sears to create the same type of response among local retailers in the smaller cities and towns of Mexico that it created in Mexico City 20 years earlier.

### *Sears Peru in 1978*

Sears Peru had four stores, all in the Lima metropolitan area: an A store in San Isidro, one B store in downtown Lima and one in the new shopping center in San Miguel, and a satellite in Miraflores. Whereas the first two stores, San Isidro and Lima Centro, were built in the mid- and late-1950s, the only expansion that took place up to 1976 was the building of the Miraflores satellite in 1966. In fact, it is interesting that

Sears went ahead with the building and opening of a major new store, San Miguel, in the mid-1970s, at a time when the revolutionary government's policies continued to be highly nationalistic, the future of the economy was very uncertain, and there were rumors that worker participation would be declared for the commercial sector as it had been for the industrial and mining sectors.

Several factors may have contributed to this decision. One may have been the good reputation that Sears enjoyed with the military government. One reason for this reputation was the Sears policy of profit sharing—a program once cited by the president of Peru as inculcating the concepts the government was seeking in advocating worker participation in industrial ownership. Thus, the government's attitude may have helped build the confidence necessary for additional investment.

In addition, in the long hiatus between the building of the Lima store and the new San Miguel store, the population of metropolitan Lima more than doubled. Several stores, competitors of Sears in a number of lines, were expanding rapidly. The decision to expand into a middle class area that was also at a major intersection in a rapidly expanding part of the metropolitan area made competitive sense. Finally, one must add the limitations on profit repatriation introduced in accord with Andean Common Market regulations. An obvious use for unremittable funds was local reinvestment.

When Sears entered Peru, its major competitor was a locally owned department store, Oeschle, which was founded in Peru around 1890. Competition grew rapidly after Sears entered the market. Oeschle expanded to several satellite stores, and by 1978 had three large department stores, including a new anchor store in a shopping center. It had credit available, although the price to the consumer was higher than that of Sears and the terms not as long. Its credit system appeared to be modeled on the Sears systems. Oeschle did not have its own repair service facilities, but referred customers to the factories.

There were two variety store chains. Monterrey, which combined variety goods with supermarkets, established its first store in 1952 and continued expanding. It had 19 stores in the Lima metropolitan area and one in the provincial city of Arequipa. A second variety chain, Tia, had four units in the Lima area and one in the provincial city to the north, Trujillo. Its major merchandise was lower-cost soft lines (clothing, linens, kitchen goods, and so forth). Neither chain offered credit or repair and service facilities.

There was also a chain of discount stores, Scala, which combined general merchandise with supermarkets. Scala had five units in the Lima-Callao area. It did not offer credit or repair and service facilities. Other competitors were the specialty shops selling clothing, appliances, kitchen goods, hardware, and so forth. There were a large number of them, many of which offered credit that was generally on more restrictive and expensive terms than Sears. Some offered repair services, whereas others referred customers to the factories. Despite the competition from a variety of stores and the economic situation, Sears had been maintaining and even increasing its penetration of the market.

Although there was little concrete information that would enable the comparison of the socioeconomic levels of the clients of various stores, one market research study did ask respondents which brands they preferred. Oeschle and Sears were chosen most often by those in the middle and upper classes; Tia, Scala, and Monterrey by those in the lower classes.[19] A 1974 Sears market research survey of customers in Sears stores provides additional indication that Sears and Oeschle drew on a similar clientele. Those holding credit cards from Oeschle or from both Oeschle and Sears had very similar income levels, the largest percentage being in the middle and upper-middle. To even the casual observer, it was obvious that Lima had neither the market size nor income distribution to develop the broad range of department stores that typifies Mexico City.

What the future held in 1978 was a question heavily dependent on the state of the Peruvian economy. There were clearly new markets to be tapped, particularly in the provincial cities. But the transport and communications problems of Peru, encompassing as it does the spine of the towering Andes, made them a significant challenge. But, more importantly, with depression and inflation reducing profits, with funds scarce internally and virtually unavailable externally, and with a return to civilian rule and new policies just around the corner, the situation was too unsettled for long-term planning.

Sears was holding its own in terms of image in Peru, but the handwriting was clearly on the wall. Its competitors, particularly for the middle and lower-middle class market, were expanding. It could not expect to maintain market position or staff morale over the long term without a dynamic, growth-oriented program. Whether or not it would adopt one would be a function of both the Sears international strategy and the recovery of the Peruvian economy.

## THE TRANSFER OF MASS MERCHANDISING TECHNOLOGY

Mass merchandising involves a series of technologies: systematic methods of creating, distributing, promoting, selling, and servicing products.[20] Over the last 90 years, Sears has developed a system of six core technologies: merchandise, operations, credit, facility development, personnel, and finance. The functions of these technology systems are briefly described in the Appendix. The transfer of these systems was the critical component in developing successful Sears operations abroad.

There are three primary means by which technology is transferred from Sears International at Sears headquarters in Chicago to the foreign corporations: training; documentation; and the direct activities of Sears International personnel. These are used in an integrated manner to transfer Sears techniques to local corporation personnel whose responsibility it is to adapt them to local circumstances.

### Training

Training is at the center of the system for transferring technology. Everyone is involved—clerks to executives. The amount of training is extensive, indicative of the importance placed by Sears on understanding its mass merchandising system. As of 1978, executives in Sears Mexico and Sears Peru had each received over 1,000 hours of training. Although this training was obviously of value to Sears, the fact that the vast majority of executives leaving Sears found employment in job-related areas indicates the importance of such training in the general job market. Thus, training in the skills of mass merchandising must be considered one of the contributions made by Sears to the economies in which it operated.

There are four basic Sears training programs: formal; on the job; ongoing; and learning visits and exchanges.

Formal training is required before persons are considered able to perform their tasks and are permanently assigned to a position. Whereas all employees, including sales personnel, receive some formal training, the most extensive course is given to those hired as management trainees. Candidates for the Management Trainee Program are usually recent university graduates. They must pass a five-hour Executive Battery Test as well as a series of interviews with Sears executives.

The *Management Trainee Program* lasts one year. The first portion consists of 31 weeks spent rotating through all the departments of a full-service store. Training manuals carefully outline what is to be taught and what skills are to be imparted by each department during this phase of the program. Throughout this period, the trainees are supervised by the department manager and are expected to learn the functions of the department and their relationship to the total operation of the store. In addition, the trainees are expected to read all the manuals pertinent to the work of that department. At the conclusion of their stays in each department, they are to write a report summarizing what they have learned about the department, which is used by the Personnel Department to evaluate their performance and potential.

Following the 31 weeks of rotation, the trainees are assigned for 21 weeks as an assistant to a division manager to learn in detail from experienced management how a selling division functions. Completion of the Management Training Program also requires passing three courses from the Sears Extension Institute (SEI): "Management of a Sears Division," "How to Merchandise a Division," and "Merchandise Administration." These courses represent a total of five person-days of study. (A copy of the Management Trainee Program can be found in Table 3.1).

Another management training program is designed for existing personnel who have shown promise in their work and

**Table 3.1. Management Training Program**

| | |
|---|---|
| Induction Training (1 week) | Central Warehouse (1 week) |
| Receiving and Marking Merchandise (2 weeks) | Fashion Center (1 week) |
| Shipments (1 week) | Assistant to Merchandise Division Manager, Soft Lines and Miscellaneous — Store (2 weeks) |
| Merchandise Control (4 weeks) | Assistant to Merchandise Division Manager of Hard Lines and Big Tickets—Store (2 weeks) |
| Client Service (4 weeks) | Assistant to Corporate Marketing Manager (2 weeks) |
| Credit (2 weeks) | Assistant to Operations Superintendent — Store (4 weeks) |
| Display — Publicity (1 week) | Assistant to Head of Division (21 weeks) |
| Personnel (2 weeks) | Inventory/January/July |
| Auditing — Payroll (1 week) | Division |
| Service Center (1 week) | |

interest in moving into management positions. These individuals generally have been with Sears for several years and may have worked their way up to the position of manager of a selling division. Because of their existing experience with Sears, the training period is only 24 weeks, divided between rotation through the various departments and planned work with the merchandise managers and selling supervisors.

The above training programs prepare people for management positions in all but two departments, Finance and Credit, both of which require specialized training. An accounting background is required of management personnel in Finance. For credit management, there is a management training program that consists of 26 weeks of rotation through different sections of the Credit Department.

Once the training program is completed and individuals assigned to jobs, they are given *on-the-job training,* the amount of which varies with the position. On-the-job training is defined as that training necessary to make persons fully productive at their jobs. It is planned work with experienced management, supplemented by the use of manuals and films. A supervisor may spend one hour per day for one and one-half or two months with a person newly assigned to a job before that person can adequately perform the necessary tasks.

*Ongoing training* is that which persons receive in order to improve their skills while they hold their jobs. It is given great emphasis by Sears, as indicated by the large number of extension courses designed for this purpose and by the fact that participation in such courses is considered in promotion. SEI has a large number of courses available for those in many different positions. Courses include "Principles of Persuasion," "Merchandise Administration," and "How to Manage a Sears Division." There are extensive courses for automobile and service technicians. These courses, given by correspondence, are available in Spanish and include examinations and homework that are graded by SEI at Sears International.

Those technical courses taken by Sears Peru technicians on staff in 1978 (Table 3.2) provide a good example of the depth and breadth of technical training available through SEI. With 61 technicians employed (not including auto mechanics), the extent of their training is obvious.

**Table 3.2. Technical Courses Taken by Sears Peru Technician**

| *Course* | *No. of Graduates* |
|---|---|
| The Automatic Washer | 6 |
| Basic Principles of Electricity | 68 |
| Understanding Tube and Transistor Circuits | 11 |
| Understanding and Using Test Instruments | 10 |
| Basic Refrigeration | 19 |
| How AC and DC Circuits Work | 22 |
| Motors and Generators | 8 |
| Basic Television | 19 |
| Small Gasoline Engines | 6 |
| Advanced Refrigeration | 13 |
| | 182 |

Executives also receive ongoing training, in addition to SEI courses, which may consist of special seminars or lectures in their particular fields, such as "New Marketing Techniques" and "Warehousing and Distribution Methods," and special management courses, such as "Management in Action," "The Effective Executive," "Executive Development," and seminars in Personnel, Operations, Credit Sales, Unit Buying Control, Service Center Management, and so forth.

The attention given to the training of sales staff indicates the importance placed on their efforts. Passing of the Mental Abilities and Temperament tests is a basic requirement for all sales personnel. Training consists of two to three days of introductory instruction in Sears operations and sales methods. This training is based on Sears International course outlines and includes several films such as "Satisfaction Guaranteed or your Money Back," "Courtesy," and "How to Be Successful in Selling." Those who sell furniture and appliances generally receive an additional one-day course in "Principles of Persuasion." After basic training, approximately one hour per week is devoted to training in new products, credit policies, sales procedures, and so forth. Product information training is often done by the buyers. In addition, there are one or two meetings per month on techniques of selling, at which Sears film strips may be shown and discussion of sales experiences and how to handle them is encouraged. Thus, by the end of the first year, the average salesperson will have received some 70 hours of training. And training in products and techniques continues throughout employment at Sears.

There are two types of *learning visits and exchanges* under which executive personnel are sent abroad. Sears Headquarters develops and conducts a series of seminars for Sears executives, including the Sears Staff School (a two-week orientation to the entire Sears system), management seminars, marketing seminars, and so forth. One component of many of the seminars is a computer simulation program, in which the participants spend one week working through a computerized simulation of one year of operations of a Sears store. This provides an invaluable means of reinforcing the interrelationships and effects of individual decisions on the system as a whole.

A second type of training abroad consists of visits to Sears Headquarters and to other Sears corporations or suppliers. The objective of this training is to bring the trainee up to date in new methods and techniques that are being developed and used in the Sears system or by Sears manufacturers. For example, operating personnel from Sears Peru and Mexico have attended the International Service Management Course in Puerto Rico and Venezuela; merchandising personnel have gone to Colombia, Mexico, Venezuela, Puerto Rico, and the United States for courses such as Basic Display and work on new systems or market research; Sears Mexico credit personnel have attended Regional Credit Seminars in Costa Rica; and personnel executives have studied personnel procedures in Puerto Rico, Venezuela, Colombia, and the United States.

### Documentation

A wide variety of materials are used within the Sears corporations. As the analysis done for this study shows, the vast majority of them are provided by Sears International with some adaptation at the local level. However, local store size and industrial development predictably play a role in the type and amount of documentation required from Sears International. Whereas the vast majority of new product ideas in Mexico were generated locally, this was not true in Peru. Mexico also had developed the capacity to provide a good portion of its technical documentation, specifications, and store blueprints and layouts.

Among the major types of documentation are manuals and other training materials, technical documentation and specifications, store blueprints and layouts, examples such as advertising mock-ups and store display ideas, and product ideas. As noted in the discussion of training, manuals that outline Sears, Roebuck's procedures serve as a basic type of training material as well as guides to everyday operations. Thus, they serve a number of functions. Among the most important of these are to set forth basic operating procedures, to act as training materials, to suggest ideas, and to provide technical specifications.

They also perform an important function by maintaining the synergism of the system through establishing common procedures. On the other hand, provision is also made for adaptation to local conditions. That this adaptation takes place is clear from a survey of basic manual use in Mexico and Peru. Given a list of some 60 basic Sears manuals, department heads in both subsidiaries were asked which they used in their department and which had been revised for local use. In Peru some 14 percent had been revised for local use, whereas in Mexico 37 percent had been so revised. In addition to those that one would expect to be revised to conform to local regulations, such as Payroll, Compensation, and Credit and Collections manuals, a number of those involving Operations and Merchandise were also revised, particularly in Mexico. Both the size of the corporation there and its geographic spread would appear to account for this.

In order to ascertain the use made of materials from Sears International compared with those from other sources, department heads in each subsidiary were asked to estimate their usage of various types of materials by source. Table 3.3 summarizes their answers. Some interesting conclusions can be drawn from this table. It is clear that Sears International was the predominant source of documentation and materials except for product ideas and store blueprints and layouts in Mexico. Almost all training materials came from Sears International, as did examples such as advertising mock-ups and store display ideas.

**Table 3.3. Sources of Documentation of Materials**

| *Documentation of Materials* | *Country* | *Local Sears Subsidiary* | *Sears International* | *Other Sears Subsidiaries* | *Local Suppliers* | *Foreign Suppliers* | *Magazines, Catalogs, etc.* |
|---|---|---|---|---|---|---|---|
| Training materials | M | | | | | | |
| | P | | | | | | |
| Technical documentation and specifications | M | | | | | | |
| | P | | | | | | |
| Store blueprints and layouts | M | | | | | | |
| | P | | | | | | |
| Examples such as advertising mock-ups, display ideas, etc. | M | | | | | | |
| | P | | | | | | |
| Product ideas | M | | | | | | |
| | P | | | | | | |

Hatched areas indicate amount of material coming from source indicated. M, Mexico; P, Peru.
*Source*: Questionnaire survey of department heads in Mexico and Peru.

The distinctions between Mexico and Peru shed light on how the role of Sears International has changed with the size of the subsidiary and the level of development of the country in which it was located. Whereas Peru continued to depend almost exclusively on Sears International for technical documentation and specifications and store blueprints and layouts, Mexico had diversified its sources. The local corporation itself was able to provide almost 50 percent of its store blueprints and layouts, which it did through its local Facility Development Staff. For technical documentation and specifications, Sears Mexico relied on local suppliers, magazines, and catalogs for over 35 percent of its needs. This ability to rely on local suppliers was even more obvious when one looked at the source of product ideas. Some 70 percent of product ideas in Sears Mexico were attributed to local suppliers—a clear indication of the extent of local development of a consumers goods industry.

One final area of interest is examples such as advertising mock-ups and store display ideas. This is the heart of marketing, and it is interesting to note that such heavy reliance for this key function was placed on Sears International by both corporations.

**Sears International Personnel**

The most important mechanism of transfer is clearly the Sears International staff, all of whom have broad experience in various aspects of mass merchandising. In addition, they have access to technical assistance, to new ideas generated in the United States and other markets, and to the experience of the entire Sears system in the United States, Canada, Spain, and Latin America. Because of their familiarity with both the Sears system and the individual conditions in each of the subsidiaries, they represent an efficient means of providing ongoing support, new ideas, and problem-solving capabilities to the subsidiary. They serve as intermediaries, translating and transferring relevant experience from the Sears system to the subsidiaries, on the one hand, and assisting individual subsidiaries in resolving problems, obtaining information, and making contacts abroad, on the other hand.

Both Sears International and Sears subsidiaries are organized by technologies, or divisions: Merchandise, Operations,

Credit, Facility Development, Personnel, and Finance (see description of functions in the Appendix). Thus, the credit manager in Sears Mexico has access to experts in the Credit Department of Sears International. Most of these experts have experience in Sears domestic operations. In addition, many have worked abroad for Sears. Thus, they are familiar both with Sears methods and with international operations.

Sears International personnel play three essential roles: supporting existing operations; assisting in solving problems; and developing and/or transferring new systems. In carrying out these functions, they spent on average 20 percent of their time working on Mexico and Peru, including trips to the countries each year. For the subsidiaries they represented a resource unavailable elsewhere at any cost, because of their detailed knowledge of the Sears system and their access to the experience embodied in it.

There are a number of ways in which Sears International personnel support existing operations of the subsidiaries. They provide ongoing review of the operations, technical assistance, new ideas, training, and problem-solving capabilities. These often enable them to spot problem areas that might not be obvious to those directly involved. For example, several years ago it became apparent to a Sears International credit manager that there were real morale and operating problems in the Credit Department in Mexico. Turnover was high, employees were discontented, and outside collectors were being used to do collection work trained staff should be able to do more efficiently. After analyzing the situation, he concluded that the problem lay in employee selection and training; so he worked with local credit personnel to strengthen the trainee program and monitored its implementation. The improved training program built a base of well-trained personnel, improved morale, and enabled the department to put staff to work on credit collection.

A major function of Sears International personnel is providing access to *technical assistance*. This is particularly important to product and source development. For example, in a survey of 50 companies that had received technical assistance from Sears Mexico and Sears Peru, 97 and 74 percent, respectively, had received some of that technical assistance from Sears International. So, even though product ideas may have been generated locally, as in Mexico, technical assistance from abroad was still needed. The assistance might take the form of manufacturing techniques; provision of samples, specifications, and drawings; arrangement of factory visits in the United States; and establishment of contact for the local manufacturer with a U.S. manufacturer from which technical assistance could be received directly. The latter was generally the case in large appliance manufacture, where Sears International had established a relationship between Mexican and Peruvian manufacturers and Whirlpool, Emerson Electric Co., Universal Rundle, Roper Co., and others. In fact, in at least one case in which a local appliance manufacturer was having difficulties in obtaining a license and technical assistance from a U.S. manufacturer, Sears International secured the relationship for him.

Technical assistance is also available through the laboratory in Sears Chicago, which will evaluate products sent up by the foreign corporations. Laboratory tests revealed safety problems with car batteries that Sears Peru was considering buying, and poor performance problems with a power capacitor in a hand tool motor. Ways in which the problems could be corrected were recommended by the lab. These reports were made available to the local manufacturers and the problems were corrected. Laboratory testing is vital in achieving the product improvements that enable the local Sears corporation to purchase the product with confidence that it meets Sears quality and safety standards.

The *provision of experts* is another form of assistance from Sears International, particularly in areas such as store construction and shopping center development, in which it would not be economical for each local corporation to develop and maintain expertise. The first Sears shopping center in Mexico was the idea of a real estate man from Sears International. Working with a local architect, Sears became a pioneer in shopping center development in Mexico. Technical assistance in design, fixturing, and construction oversight was provided by Sears International. Other major shopping centers built in Mexico have also had foreign technical assistance.

An indication of what can be involved in terms of time was provided by the expansion of the Fashion Merchandise Distribution Center in Mexico. The expansion was coordinated by Sears International Operations but also involved Merchandise and Facility Development in both Chicago and Mexico City. Some six trips to Mexico by Sears International personnel and over 400 person-hours at work in Chicago were involved in coordinating the design and construction of the new Center.

Finally, Sears International is equipped to provide *training and materials* for ongoing operations in areas in which it would not be efficient for the local corporation to develop the capability, such as advanced market research and data-processing techniques. As in many other areas, when developed for a large number of corporations, the cost can be spread out and it becomes economical to provide the resources.

*Ongoing review* by Sears International personnel not only results in spotting problems, but provides the in-depth knowledge necessary to solve them—another important function of Sears International. For example, many of the corporations have, at one time or another, had to deal with the problems of purchasing, pricing, providing credit, and selling in highly inflationary economies. An important function of Sears International is to transmit the results of their experience to other corporations facing similar problems. For example, Sears Peru received assistance in adjusting credit policies to a highly inflationary economy—assistance based on Sears experience in Brazil and other countries. Such assistance played a role in helping Sears maintain reasonable credit terms during the inflation and recession that Peru suffered during the late 1970s.

The introduction of the Seasonal Unit Sales Plan (SUSP) into Peru provides another example. This is a system of accounting for monthly sales by units rather than dollars or soles (Peruvian currency). In inflationary economies, the ordering of merchandise on the basis of previous local currency-denominated amounts is obviously unsatisfactory, as they do not provide an accurate guide as to the number of units required. SUSP was developed to avoid stockouts and enable more efficient planning by both the buyer and manufacturer. Responsible for the introduction of SUSP were merchandising managers from Sears International. The estimated time for complete introduction and implementation was one year, or two buying seasons. Several visits were made by the merchandising managers involved in order to train local personnel in the system and oversee its implementation.

Careful review by Sears International personnel can result in savings through applying experience from elsewhere to avoid future problems. For example, Sears U.S. had a difficult and costly time converting its service manuals to microfiche because of the poor quality of the paper on which they were printed (they were published before the problems of conversion to microfiche were considered). The lessons of this experience were relayed to Sears Mexico with the recommendation that they begin to upgrade the quality of paper on which service manuals are printed in order to avoid the same problem when, five to ten years hence, the volume of service material used would be such as to make conversion economical.

Another example of Sears International's role as a *problem solver* involves the production of fixtures for new stores. In many of the countries in which Sears has operated, local manufacturers have been unfamiliar with the techniques used in making the round steel tubing required for fixtures for new or remodeled stores. Thus, International Fixturing developed a demonstration model that graphically displayed the steps necessary to make the tubing. It has been used in a number of countries by local manufacturers who now produce the required tubing not only for Sears but for other stores as well.

*Developing new systems* is another important function of Sears International. It is a particularly appropriate one because of International's worldwide access to new but tested mass mechandising ideas and its ability to spread the cost of new system development over a large number of corporations. Perhaps the best example of new system development was the introduction of a data-processing system into the Mexican corporation. With the growth of credit accounts in Mexico and the concomitant increase in paperwork, it became necessary to mechanize the system. While the new system developed by Sears International would first be applied to credit balances, providing information on average daily credit balances rather than end-of-previous-billing-period balances, it would also be able to handle such other data needs as accounts payable, social security and benefits payments, merchandise reports, and so forth.

Included in the development of the system by Sears International Data Processing were the programming of the computer, writing of manuals for its use, and development of training courses for local personnel. Sears International agreed to provide a trainer to conduct training courses in Mexico and to oversee the conversion of the new system. Total time spent by Sears International to develop the system was estimated at two and one-half person-years.

The introduction of this system provides a good example of the economies of scale available via Sears International. Although the system was established for Mexico, it was designed with all of the corporations in mind, so that once tested in Mexico, it could be introduced into the others when it would be cost effective to do so. Thus, development costs could be spread over a larger base than only Mexico would provide. Adapting it for Peru, for example, would require an estimated 10 percent of the time cost entailed in developing the original system.

Sears International's assistance taking place, as it does, day after day, year after year, represents a significant inflow of knowledge and experience to the Sears subsidiaries. This is not to say that all of the recommendations are great cost savers or that oversight from Chicago may not at times be viewed as excessive and intruding. Nor must one be blind to the fact that the type and amount of assistance change as the subsidiaries grow in size and experience. Certainly, less was provided in 1978 from abroad in the area of product and source development and more done locally than was the case 15 or 20 years previously. On the other hand, there was a tremendous inflow of assistance in the area of data processing. Such changes are natural, as one of the major roles of Sears International is to transfer knowledge and train local personnel in its use. As long as there are innovations in mass merchandising, a major function of Sears International will be to transfer them.

## MASS MERCHANDISING AND DEVELOPMENT

### Introduction

The goal in any economy should be to achieve self-generating growth—to turn a static situation into a dynamic one. The technology of mass merchandising is particularly suited to contributing to this, because the concept on which it is based is that of market growth.

Mass merchandising introduces a multiplier into the economic equation. It acts in two basic ways: to increase the size of the market, and to rationalize the production and distribution of goods. The market is increased by reducing the cost of goods, promoting their sales, providing credit for their purchase, and producing new products that the consumer wants and needs. Enlarging the market increases employment in the sector producing the goods, as well as in the service sector itself. This increased purchasing power translates into an even larger market through the multiplier effect.

Mass merchandising rationalizes production by reducing waste through introducing efficiencies into the production and distribution system. Transportation, inventory control, warehousing, and consumer research are examples of areas in which efficient operation can be translated into reduced costs and resource wastage. Mass merchandising also rationalizes production through the creation of larger markets that enable economies of scale to be achieved in such areas as production, distribution, credit, and advertising. The larger the market, the lower are the unit costs assigned to individual products and therefore the more that can be sold. Here, too, the multiplier is at work.

The following sections will use the operations of Sears, Roebuck and Co. in Mexico and Peru to illustrate the impact of mass merchandising on economic development. Five operational areas will be analyzed: product and source development, product marketing, behind-the-scenes operations, credit, and training. Also examined will be the adoption of Sears technologies by others in the retail sector because of the ease of identifying and employing them.

### Product and Source Development

The development of manufacturing sources is one of the

major, if unintentional, contributions of mass merchandising to economic development. Sears entered both the Mexican and Peruvian markets planning to supply a good proportion of merchandise from abroad, particularly large appliances and more sophisticated manufactured goods. But, using its U.S. experience, it was able rapidly to adjust to national import substitution legislation when it was enacted. And in some cases, as with appliances in Peru, it assisted local production before import substitution requirements greatly reduced external supplies. In both Mexico and Peru, Sears opened its first store with approximately 50 percent of the merchandise purchased locally. Within ten years of opening, local purchases in both countries were some 80 percent of the products sold. In 1978 the figure was 99.8 percent for Mexico and 97.9 percent for Peru. In Peru an additional 1.6 percent was purchased from within the Andean Common Market, primarily appliances from Colombia. It took approximately 24 years in Mexico and 22 years in Peru for local sourcing of Sears merchandise virtually to double. How this occurred is the story of product and source development.

A detailed picture of what Sears does and how it does it was provided from a questionnaire for this study, which asked for 25 to 30 examples of product and source development. Sears Mexico listed 29 different companies making products ranging from stoves, refrigerators, and dishwashers to pantyhose, exercise equipment, citizen band radios, and a variety of types of clothing. The average number of person-days spent up to 1978 by Sears Mexico merchandise managers in providing technical assistance to these companies was 38.5 per product. Four of the companies were put in contact with U.S. manufacturers, often Sears U.S. suppliers, such as Whirlpool, from whom they could receive technical assistance directly. Sears International personnel were involved in providing technical assistance in 28 of 29 of the cases cited.

The figures differed somewhat for Peru. It was a smaller market and farther from Sears International headquarters, so the economics of direct assistance were different than in Mexico. Peru was also in a different economic situation, which may have meant that less new product development was called for. Listed were 26 companies that received technical assistance for

34 different products ranging from refrigerators, hot water heaters, and stoves to pressure cookers, furniture, kitchen cabinets, suitcases, and some clothing. An average of nine person-days per product had been spent by Sears Peru merchandise managers to date in providing the technical assistance. Two of the companies were put in contact with U.S. manufacturers. Sears International assistance was provided for 27 of the 34 products.

But the most interesting and telling fact was that, of the products for which it has provided technical assistance, Sears bought an average of only 30 percent in Mexico and 25 percent in Peru of total production by the supplier. In other words, the improvements that Sears promoted redounded to the benefit not only of Sears but of its competitors as well who also purchased and sold them. This was a result of both a business decision and the market. Serving as sole outlet for a manufacturer raises problems of dependency that Sears states it would prefer to avoid. Market size and economies of scale also argue for larger production runs than Sears alone could purchase. In fact, in Peru, Sears even encouraged an appliance manufacturer to set up its own outlets, in competition with Sears, in order to achieve a market size that would allow the desired economies of scale in production.

One way of dealing with the obvious competitive problems caused by this approach is to have an "exclusive" made only for Sears, which may involve differences from the basic product in style, color, or performance. The manufacturer can make a basic production run, varying only those aspects required to obtain an exclusive model. Where such an approach is possible, Sears has followed it, as in Mexico. In Peru, where retail brands have been made prohibitively expensive by a special tax, Sears has not had exclusives.

Sears was more heavily involved in its early years in helping to establish local manufacturers than in the late 1970s, although some source development was still being done, particularly in Peru. The reason for this is clear. Local manufacturers capable of producing quality electrical appliances, for example, scarcely existed when Sears entered Mexico and Peru. To obtain locally made refrigerators, stoves, water heaters, and other appliances, Sears had to help create the companies as well as provide a wide variety of technical assistance.

The list of assistance provided to a major appliance manufacturer in Mexico was a good example of the type of assistance available: production planning, drawings and specifications, development of manufacturing facilities, introduction of advance administration and accounting procedures (including standard costs), assistance in importing raw materials from Sears U.S. sources when they were not available locally, visits to the factories of U.S. Sears sources, and assistance in exporting and contact to develop a licensing arrangement. In the beginning Sears had a buying contract with the company guaranteeing purchase of 80 percent of its production. By 1978 the company had grown to the point that the Sears purchasing level was less than 10 percent. However, Sears was assisting in developing export markets for its products through Sears U.S. Export tests were taking place and it was hoped that within a few years, it would be exporting some 15 percent of its production of refrigerators.

Crucial to this type of manufacturing assistance, particularly as the source or product was being developed, was the basic buying contract, as is indicated in the following quotation from a Peruvian appliance manufacturer, INRESA:

> . . . said contract committed Sears to buy all of her refrigerator needs from INRESA in the sizes and types produced . . . taking the risk that the product might not be of the same quality that had been provided to her clients (through imports). There is no doubt but that without a contract of this type INRESA would have had many more difficulties than those which it had in selling its product in a highly competitive market given the models which were freely imported into the country.[21]

Sears facilitated the licensing contract that INRESA obtained with Whirlpool.

An example in another less technical field is provided by the relationship between Compas, a men's clothing manufacturer in Peru, and Sears. At the same time Sears established its first store in Lima, Compas began with two sewing machines and five operators. By 1976 it had grown to almost 600 employees and had a license with Jockey for part of its production. According to the general manager of Compas, the relationship with Sears had been most fruitful:

> Technicians of Compas, sponsored by Sears, traveled to Chicago, were put in contact with important factories, which led to the manufacture of the first suit jackets at an industrial level even though on a minimum scale.
>
> During the following years, the demand of Sears Roebuck del Peru, S.A. grew, continuing to open new stores, and the small factory, Compas Industrial S.A. grew as well thanks to the support of his client. It grew in machinery, in volume of production and diversification of the same, and, what is most important, it constantly developed in the area of improved quality, since Sears had the ability, not only to pressure in order to obtain this, but to provide the support of ideas, techniques and all that was necessary to improve the garments more each day.[22]

In addition to helping establish manufactures, Sears has introduced numerous new products to the local markets in Mexico and Peru. This means finding manufacturers to produce the product, and often providing designs, specifications, and samples; it may also call for assistance in manufacturing techniques. Lower price and better quality were the reasons for the technical assistance given in developing a polyester-cotton blend for use in underwear in Mexico. When the peso was devalued in 1976, Mexican cotton prices soared as the manufacturers sold at the world rather than the peso price. This drove up the price of cotton underwear. So, Sears buyers set out to develop a new product that would be less expensive and have improved durability. Their greatest problem lay in convincing the manufacturer to change. They finally succeeded and the new line was introduced in 1977, causing the greatest growth in the manufacturer's history. Sears bought only a small percentage of total production.

Product and source development is crucial technology in the Sears mass merchandising system. The unique combination of the knowledge of the merchandise manager in the local corporation and Sears International merchandise managers and their access to worldwide market trends, technical information, and know-how makes it possible. It is one of the mass merchandising techniques that Sears employees say has not been widely adopted by Sears competitors. Certainly, it is a technology that redounds to the benefit of Sears competitors, because the results of Sears efforts are sooner or later in the general marketplace where they can be bought or adapted by others.

The contributions of product and source development to economic development are fairly clear. Both directly and indirectly they help to mobilize latent economic resources and put them to productive use. They work directly by taking an idea, like the Popcycle sandle, and turning it into jobs and production. They provide assistance in organizing production to make it more efficient, as was done with introduction of standard costs to an appliance manufacturer in Mexico. They provide direct technical assistance in the form of designs, specifications, and manufacturing techniques. They also help to guarantee the safety of products sold by Sears.

Product and source development policies and procedures also have an indirect impact on mobilizing latent economic resources. The drive to develop lower cost products, as in developing and introducing modular kitchen cabinets, helps to increase demand. Insisting on quality products that are safe helps to ensure that the demand levels and therefore production levels will be maintained. Finally, the entire spectrum of relationships between the merchandise manager, the buying system, and the local entrepreneur might be viewed as an exercise in on-the-job training of local entrepreneurs. All of these activities, undertaken to create an efficient and profitable store, also have direct impact in making the local economy a dynamic and growing one.

## Marketing the Product

Product and source development have obvious, direct effects on economic development, but they are totally dependent on effective marketing of the products. Well-located stores mean more customers. Good promotion of merchandise means more sales and lower inventory costs. Good quality products and repair facilities translate into future sales. All of these are ultimately reflected in creation of jobs and efficient use of resources, be they human, financial, or material. Store location is an area in which Sears has pioneered in Latin America, based on its experience in the United States. When the first store was built in Lima, the location selected was far from the downtown shopping area but near a number of well-to-do districts. Sears may have been the first store to locate there, but it was soon followed by others who noted its success. The area became a major shopping center. Sears has led the way in shopping center development in many areas of Latin America, including Mexico.

A successful shopping center means much more than profitable operations for the anchor stores like Sears. It means life or death for the many small, local shopkeepers located in it. For this reason, well-trained management and the ability to draw on experience from elsewhere are vital. The major shopping centers in Mexico, both those of Sears and of others, have had foreign assistance in their design. Sears has taken management responsibility for the shopping centers in which it is involved in both Mexico and Peru.

Product promotion helps to increase sales by acquainting the consumer with the product and ensuring consumer satisfaction. Of importance in the Sears system are identification of what the consumer wants, assurance that price and quality are comparable with those of other stores, guarantee of quality merchandise, availability of repair and servicing, courteous staff, and, of course, advertising.

Sears famous guarantee, "satisfaction guaranteed or your money back," is an integral component of product promotion, for it increases customer confidence in the products purchased. The importance of this approach to successful merchandising is indicated by the fact that some of Sears competitors have adopted it.

An example of how successful a well-run and well-timed promotional campaign can be in increasing sales was that for disco clothes based on those in the movie "Saturday Night Fever." Having seen how popular the film was in the United

States, the Mexican merchandise manager had Sears International obtain the rights to produce "Saturday Night Fever" pants, suits, shoes, and T-shirts in Mexico. They were ready for the Mexico City opening of the show. Promoted in the local media, the clothes were located at a central point in the store, with "Saturday Night Fever" music blaring to attract attention. The result was an over 100 percent increase over the previous year in sales of pants, suits, shoes, and T-shirts.

An indication of the importance of repair and service can be found in two studies of consumer satisfaction done several years ago by Sears Peru. The first study indicated that some 87 percent of the customers would buy their appliances at Sears again based on their experience with repair and service. However, some problems were noted, and efforts were made to correct them. The same study done the following year found that 96.5 percent would buy again at Sears. Clearly, attention to this area is important to sales.

A good reputation in repair and service is believed to be one of the important factors in making Sears a major retailer of appliances. None of Sears large department stores competitors had their own repair facilities; customers were referred to the factory for service. The same was true for most of the specialty shops. Sears had both service centers and service technicians who would visit the home to repair a large appliance. In Mexico, for example, there were 285 technicians and 190 service vehicles that made an average of 250,000 calls per year. In both Mexico and Peru, the cost and time of Sears repairs were estimated to average less than those of other facilities available. In addition, Sears had credit available for repairs, which most of its competitors did not. It also guaranteed its work for 90 days and replacement parts for one year.

## Behind-the-Scenes Operations

Those operations involved primarily with ensuring efficient operation of the store and corporation, rather than with direct development or promotion of products, might be classified as behind-the-scenes operations. Included would be such areas as financing, inventory, warehousing and transportation, construction and operation of buildings, and, of course, training and motivation of personnel.

The effect of the behind-the-scenes operations on economic development is often overlooked, probably because the individual operations themselves are seldom large or glamorous. Yet, a cursory glance at any book on productivity will indicate that it is the combination of careful concern for ways, large or small, to reduce costs and improve efficiency along with the training and motivation of personnel that is key to maintaining and improving productivity. Given the shortage of resources in developing countries, it is precisely the increasing of individual productivity that is vital in the development process. It should also be recognized that reducing operating costs is one of the major ways of maintaining prices at the level at which large portions of the population can afford them. Thus, behind-the-scenes operations contribute to the economic development through improving the efficiency of resource use and, indirectly, through helping to reduce product costs.

An important aspect of the Sears system is planned buying. It is undertaken in order to maximize efficiency within the system through relating projected sales to merchandise needs. However, it also has an important impact on the manufacturers in that it enables them to plan production, purchasing, and storage efficiently. Many of the competitors of Sears simply buy what is available from the manufacturer or place an order for a large item, such as furniture, once the customer has requested it. This causes inefficiencies in the manufacturing process; either set-up and -down times are increased by making small production runs, or storage costs are increased to keep merchandise in inventory. In addition, raw material prices may be higher because bulk purchases cannot be made. The importance of Sears buying contracts was discussed by INRESA, the Peruvian appliance manufacturer:

> The contract . . . that Sears has with local industrialists . . . guarantees the purchase of a determined volume of products which therefore can be produced in the pre-established quantities which has an advantageous influence on the costs of production, allowing maintenance of low stock of material as well as of finished products.[23]

Thus, the detailed process of planning and projecting sales not only aids the mass merchandiser, but has an important impact in promoting more efficient production planning and inventory control in the local manufacturing sector, with the concomitant savings these imply.

The introduction of the SUSP provides an example of how a system, introduced by Sears International in order to make the local Sears operations more efficient, has a similar effect for suppliers to Sears. By providing a means for more accurately predicting purchasing needs in softwares, stockouts and inventory hangovers can be more easily avoided by both Sears and its manufacturers. Such avoidance, in turn, means more efficient management of time and money.

Improvements in inventory control have also been obtained through introduction by Sears International of an Inventory Recovery Manual to all international corporations. It is designed to help avoid stock loss shortages—those shortages discovered at inventory time when inventory counts are compared with sales. Within one year of being put into use, inventory recoveries corporationwide improved.

Examples of efficiencies introduced by Sears in building construction abound. Sears sprinkler systems were introduced. The efficiency, of course, is lower insurance premiums. Sears International has given a great deal of attention to energy-savings devices. A new form of light, Ultralumin, was being introduced, which results in 10 to 20 percent savings in electricity over conventional lighting. It also improves the customer's vision of depth, color, and texture of the merchandise. Where economical, air conditioning-monitoring devices were being installed to control peak loads and thereby reduce electricity usage. In addition, new manuals were sent to all corporations with instructions on how to reduce energy use, such as turning off air conditioning one-half hour before store closing time. The store will remain cool enough through closing time while energy will be saved. Capacitors were also installed to reduce electricity consumption. In Peru such installation resulted in savings of $18 to $20,000 in the first year.

Other energy-saving devices were tested in various Sears International stores. These included solar panels for heating the large amounts of hot water used in the stores and sky lights for lighting storage areas instead of the traditional electric lights. These and other devices would be introduced where economical to do so, once they have been well tested.

The key to Sears ability effectively to introduce these efficiencies in behind-the-scenes operations is a well-trained and motivated staff and access to innovations, be they ideas or new mechanical devices, which have been tested and are known to fit into the ongoing system.

## Credit

Credit has a number of roles to play in economic development. Among those that are relevant to a consideration of the impact of mass merchandising are its roles as an extender of the middle class, as a multiplier and accelerator of economic growth, as a promoter of industry, particularly durable goods industries, and as a stabilizer in times of economic downturn.

### *Extender of the Middle Class*

The growth of a significant middle class has long been recognized as a symbol of economic development. Access to the items considered necessary to lead a comfortable life is the goal of most people. Providing many of these goods is what mass merchandising is all about. Determining the type of product needed or desired, producing it at an affordable price, and providing guaranteed quality and service are important factors in making goods widely available. But unless credit can be obtained, the only ones able to buy them are those with cash or savings. By making credit available, those with limited means can purchase as well. Thus, credit acts as an extender of the middle class, providing access to the goods and life-style of that class before one has saved enough to pay the entry price in cash. By so doing, credit acts as a type of forced savings, directly applied to desired goods and enabling their purchase long before savings would do so.

Integral to the concept of mass merchandising is the availability of credit to the majority of those who shop in the store. According to Sears credit managers, there were no set rules for the granting of credit. Rather, the decision was based on a series of factors among which are the applicant's income, expenses, and credit reputation. Basically, as the Mexican credit manager explained, "It is a question of judgment based on experience." Will the applicant be able to repay on time?

Who are the credit customers at Sears? Is a significant portion of them from the middle class? Although extensive and completely comparable information was not available, some basic conclusions can be drawn. An idea can be obtained by comparing monthly income levels of Sears credit clients with those of the population of Mexico City. In 1975, 63 percent of Sears credit clients were drawn from income levels representing approximately 79 percent of the population of Mexico City.

Another interesting perspective is provided by the zonal credit clients in Mexico (the zone in those Sears stores outside the Federal District that encompasses Mexico City). A comparison of the average income of credit customers from Sears stores in three different cities for which information was available with that of those in Mexico City indicates that zonal credit clients had incomes significantly below those in Mexico City. The reason for this is fairly clear when one looks at the socioeconomic data on the Mexican population. With the Mexican population divided into three socioeconomic categories, the number of individuals in the upper two categories in Mexico City was 24 percent greater than the number in those categories in the three cities used for the comparison.[24] A second factor is the greater availability in Mexico City of other purchasing sources, such as discount houses. With a lack of access to such stores in many cities of the interior and with lower average income levels, it is logical that credit customers of Sears in the zone would have lower average incomes. Given these lower income levels and Sears expansion, which was taking place primarily outside Mexico City, it could be expected that the income level of the average Sears credit customer on an absolute basis would fall over the next few years. This is another indication that credit could and was being made available on a broad basis.

Comparison with competitors in the area of credit is difficult to make, but some information does exist. That Sears

Mexico was already clearly placed in the middle class segment of the Mexico City department store market by 1969 is clear from a random sample study done by a market information organization. Respondents were asked whether they had credit accounts in any of several stores, two of which were considered to be oriented toward the lower-middle and lower classes. When arrayed by income levels, the majority of credit customers at Sears fell well between those with credit from the stores considered to be upper-middle and upper class and those considered to be lower-middle and lower class. Credit accounts also appeared to be more common among Sears shoppers, probably owing to the larger volume of sales of appliances and furniture. Whereas over 60 percent of the respondents held Sears credit cards, the comparable figures for credit cards from other stores were 45 percent and below.

Information comparable with that for Mexico was not available for Peru. However, some indications of the breadth of the credit access could be obtained. A Sears market research study done in 1973 broke Sears credit customers into four classes. Sixty-nine percent were found to come from the middle class, 8 percent from the lower income group, and 17 percent from the upper income levels (6 percent did not reply to the question). Income levels of credit card holders at Sears and its competitors were similar in all respects but one: The competitors tended to reach to a lower income level than Sears. Whereas approximately 70 percent of Sears' credit customers earned between $230 and $807 per month, some 80 percent of credit card holders from Sears competitors earned between $115 and $807 per month. Thus, although Sears introduced broad credit into Peru, others had taken it up and were making credit available to the middle and lower-middle classes.

It should be noted that the primary customer, and therefore credit client, of department stores in Mexico and Peru when Sears entered these countries was a member of the upper class, the "elite carriage trade."[25] When Sears introduced credit as part of its mass merchandising approach and made it broadly available to Sears customers, the other stores followed suit. Thus, in both Mexico and Peru, Sears appears to have had an impact both as a direct result of its credit program and by stimulating its competition to grant credit. In addition, in Mexico it could be expected that the planned corporate expansion would result in a significant increase in credit availability to the lower and lower middle income groups, particularly outside Mexico City.

*Multiplier and Accelerator*

The role of credit as a multiplier and accelerator of economic growth has been widely discussed in theoretical literature. The basic thesis is that an increase in consumer installment credit leads to higher current income, since basic income is expanded by the amount of credit available. This expanded buying power is reflected in increased consumption, which then has a multiplier effect on employment, since more goods must be produced to meet increased demand. Expanded employment means even more buying power, and the multiplier process, once induced, goes on and on. At some point in this process, if the increase in credit outstanding is large enough, one can expect that it will induce an increase in business investment: the accelerator effect. This increased investment should start its own chain of multiplier effects.[26] Thus, both the multiplier and accelerator, induced by credit availability, act to promote employment and investment.

It should be noted that the multiplier and accelerator will have maximum impact in creating growth and employment when the goods being sold are manufactured within the country. Clearly, the greater the numbers to whom credit is available for purchase of locally manufactured goods, the larger the effect of the multiplier.

To analyze the impact of the multiplier, one must look at both the numbers receiving credit and the amount of credit made available. In both Mexico and Peru, the number of Sears credit accounts increased at a greater rate than the growth of the population served.[27] Whereas the urban areas of Mexico grew a little over threefold between 1953 and 1977, the number of Sears credit accounts increased almost sixfold.[28] A more accurate guide may be the amount of credit made available, since an individual can hold two accounts (for example, revolving credit and installment plan), and some of the increase in accounts probably comes from individuals opening a second credit account. The amount of credit made available by Sears Mexico during this same period increased 25-fold, 5 times the rate of increase in the consumer price index.

In Peru figures on the number of credit accounts were available only as of 1966. Between this time and 1977, the number of active credit accounts increased 2.3-fold, while the population of the Lima metropolitan area increased 1.5-fold. The first year for which credit sales information in Peru is available is 1962. While a 7-fold increase took place in the consumer price index, a 19-fold increase occurred in the amount of credit made available by Sears.[29] It is clear that the Sears credit program was responsible for a large increase in buying power in both countries, directed primarily toward the middle class (and upper-middle class in Peru).

As noted above, Sears introduced the concept of making credit broadly available, an approach that was then adopted by its competitors. Thus, Sears has had a double effect in stimulating the multiplier and accelerator: directly through the increased purchasing its own credit program promoted, and indirectly through stimulating credit availability by its competitors.

The synergistic effects of the interrelationship between credit and other technologies of mass merchandising should not be overlooked. Primary among them is the efficiency introduced into the process by the link created between consumer research, product and source development, credit availability, and product marketing. The more direct the information flow between the provider of credit, the seller of goods, and the producer of goods, the more efficiently the accelerator will work. Those in business need accurate information on how much to produce and when to produce it. The kind of direct link that exists between Sears buyers and their sources provides exactly this type of information, serving to make the entire process more efficient, thereby freeing the resources for application elsewhere.

Thus, Sears technologies play two roles in promoting the multiplier and accelerator of economic growth: first, by making credit broadly available, and second, by making the growth process more efficient through the linking of consumer needs and desires directly to the producer.

*Industrial Promotion*

The durable consumer goods industry has been recognized as playing a "leading sector" role in carrying along the general economic advance of the United States during the first half of the twentieth century.[30] Most development theorists have looked to it to play a similar role in developing countries. The importance of this industry resides in several factors. It is a primary mechanism for the development of industrial, managerial, and technical skills. It creates backward linkages in terms of supply of inputs and forward linkages in distribution, marketing, and servicing. Relevant to the discussion here is the fact that adequate consumer buying power is crucial to the development of a durable consumer goods industry. And much of that buying power, especially in a developing country, is provided by credit.

Sears has been a major marketer of consumer durables in both Mexico and Peru. Sales of large and small appliances, furniture, and home improvements in 1978 made up approximately 48 percent of total sales of both corporations. Sears has been known for its consumer durables, as its competitors have been known for their soft lines (clothing, toys, accessories, and small specialty items). This was borne out in the 1969 market study of retailers in Mexico City noted above; the Sears reputation in appliances and furniture was stronger than that of any of its competitors.

An analysis of the use of the two types of credit available from Sears also indicates the importance of credit to consumer durable sales. Installment credit is the type used for the purchase of large items such as furniture, appliances, and home improvements. Whereas between 45 and 50 percent of all sales in both corporations fell in these categories, nearly two-thirds of all credit was on the installment plan. Clearly, the major source of purchasing power for appliances, furniture, and home improvements was installment credit.

Can one trace the effect of this credit availability on production and employment? Doing so is obviously difficult, but by using credit sales and employment estimates for source factories made by Sears personnel responsible for credit and product and source development, a rough estimate can be made. In the appliance, furniture, and home improvement areas in Mexico, it could be estimated that there would be a loss of almost 1,000 factory jobs just among persons producing these items for sale in Sears were credit not available.[31] This is based on Sears experts' estimate that 70 percent of sales of these products would be lost if credit were not available. This direct loss of jobs would be the primary impact; obviously, there are secondary and tertiary employment impacts among suppliers, transporters, and in sales itself, which are not even considered in this rough figure. The role that Sears product and source development efforts have played in assisting the creation and growth of local industries is discussed above. However, it should be noted that there is an integral link in development terms between such efforts and credit. Credit can be made available for the purchase of imported or locally produced goods, but the employment and industrial development impact of credit availability can occur only if the goods are produced locally. Thus, the Sears role in product and source development has been a vital factor in making credit availability a promoter of industry and a multiplier and accelerator of economic growth.

### *Economic Stabilizer*

The role of credit in economic stability is one that has been actively debated, particularly in macroeconomic terms. There seems to be agreement that consumer credit, if promoted during a time of economic downturn, can help to stabilize the situation by creating more buying power than might otherwise exist. In this context, the 1978 situation in Peru provided a very interesting example. As noted above, Peru was suffering both a recession and virulent inflation, which meant a drop in real purchasing power in the Lima area of over 41 percent in the five years between 1973 and 1977.[32] Gross domestic product declined in both 1977 and 1978.

Purchasing power was squeezed even more in July 1978 when the government lifted the last subsidies on basic consumer products causing rapid price rises while wage increases were limited. Reaction to this restriction of buying power was quickly evident at Sears: Sales did not rise by as much as they had in previous years. There was an increase in the amount of revolving credit, usually used for clothing and smaller goods, compared with installment credit, presumably because of concern over future earnings.

A comparison of credit charges in Peru in late 1978 indicated that those of Sears were below almost every one of its department store and specialty shop competitors. Whereas most of its competitors required a down payment, Sears had not. In response to the difficult economic situation, it began to ask for one, but the client not willing or able to provide it was not refused credit as a result. In addition, when the government authorized an increase in the amount that stores could charge credit customers, Sears decided not to raise its credit rates. As the manager of Sears Credit pointed out, sales were slow and they preferred to obtain new clients.

Although such policies are clearly believed to be in the interest of Sears, they also have an important impact on the economy. The maintenance of purchasing power is of crucial importance in times of recession, for reduced purchasing power leads to reduced employment. Whereas credit acts as a multiplier in promoting economic growth, its restriction can have the opposite effect in time of economic downturn. Thus, any reduction in credit availability, as would be caused by charging higher rates, reduces purchasing power. Hence, reduction in credit can create a downward spiral just as increased availability of credit can create an upward one.

**Training**

Training is done in order to improve the efficiency of the employee on the job. Therefore, the amount of time devoted by the company to training serves as a guide to the importance given it. Of equal import in a study of the economic impact of mass merchandising are who receives the training and the value of the training to the society at large. A detailed analysis of the training received by Sears employees in Mexico and Peru sheds significant light on these issues.

Virtually all of the training was being given to nationals of the countries involved. U.S. citizens make up 1 percent of Sears Mexico executive personnel and 4 percent of Sears Peru executive personnel. There were even fewer third country nationals, usually from other Latin American countries—0.7 percent in Mexico of executive employees and 1 percent in Peru. There were no U.S. personnel employed below the executive level.

One method of judging the value of the training to society is the use to which employees can put it in obtaining other jobs. Presumably, experienced executives leaving Sears for employment in areas related to the jobs they previously performed were being hired at least partially because of their experience and training. An analysis of the executive personnel who have left both corporations since 1970 indicates that in Mexico 98 percent of those who left voluntarily for other jobs took them in areas related to the work they were doing at Sears. The comparable figure for Peru is 80 percent.

In the period of 1970–77, Sears Mexico trained over twice as many management trainees, relative to its number of executive personnel, as Sears Peru. But Sears Mexico also lost an average of 50 percent of its management trainees, whereas the loss in Sears Peru was minimal. Over half of this number were let go by the corporation itself, primarily for unsatisfactory work performance. This led to a decision by the Personnel Department that, among other things, its selection procedures must be improved. The other half left voluntarily; 43 percent giving "better opportunities" elsewhere as the reason and 29 percent citing "lack of satisfaction with their salary." Clearly, in a growing economy where other opportunities are available, as in Mexico, Sears training would appear to be viewed as useful in the job market.

Table 3.4 summarizes the amount of training that Sears had invested in the group of executive personnel employed in Mexico and Peru in 1978. The average number of hours of training per executive in Mexico was 1,218; in Peru it was 1,369. The figures clearly indicate that the largest portion of this training took place during the management training period. However, Sears commitment to ongoing training is illustrated when one sees that executives on staff in Mexico and Peru received an average of 45 and 27 hours of formal training in their respective countries and 23 and 58 hours, respectively, of formal training abroad.

**TABLE 3.4. Formal Training: Sears Mexico and Peru Executive Personnel**

| | *Mexico* | *Peru* |
|---|---|---|
| Number of executive personnel | 263 | 72 |
| Management training[a] | | |
| Management trainee (1 year/2,080 hr) | 91,520 | 40,560[b] |
| Management training for existing employees (24 weeks/960 hr) | 210,240 | 49,920 |
| Ongoing training[c] | | |
| Locally run courses | 7,709 | 1,228 |
| Correspondence courses (postmanagement training) | 4,218 | 720 |
| Total training in country | 313,687 | 92,428 |
| Training abroad | 6,680 | 4,160 |
| Total training (hr) | 320,367 | 96,588 |
| Total training (hr) employee | 1,218 | 1,342 |

[a]The hours of management training are based on corporate records indicating the number of current employees who were management trainees and the number who received management training as existing employees. The Credit Training Program is counted as 26 weeks or 1,040 hr.

[b]Based on 19 management trainees and 1 credit trainee for Peru.

[c]All Sears executives were asked to fill out work histories that listed the training courses they had taken as well as work-related travel abroad. The response rate was 72% in Mexico and 85% in Peru. The figures for ongoing training and training abroad based on these work histories were adjusted, using the response rates as the basis for projecting training of all executive employees.

The discrepancy in average course hours between Mexico and Peru would appear to be accounted for by two factors. The first is the size of the local operations, which would make it feasible for the Mexican corporation to offer locally some courses for which Peruvian executives might be sent abroad to other Sears corporations or to Sears International. The second factor may well be the lower turnover rate of executive personnel in Peru. Whereas Mexico averaged a loss of 7 percent of its executives per year for the previous eight years (for all reasons), Peru averaged only 3 percent. Having to train larger numbers of new personnel might mean that Mexico relies more heavily on formal courses while Peru can rely on on-the-job training and experience.

Technical training is an area receiving increasing importance in the developing countries. Sears, in order to meet its commitment to repair and service the products it sells, has provided extensive technical training. All technicians had to pass a test in basic mechanics before being accepted. They were then assigned to work with an experienced technician for two to three weeks for apprenticeship training. All technicians were expected to take a series of Sears correspondence courses such as "Basic Electronics," "Black and White TV," "Basic Motors," and "Basic Refrigeration." As noted earlier, in the previous eight years, Sears Peru technicians graduated from 182 technical courses offered by SEI. Whenever a new line was introduced, the technicians would spend two to three days in the factory learning how to repair it. There were also special courses, such as "Techniques of TV," given by local technical schools to which Sears would send some of its technicians.

In conclusion, Sears as a corporation is obviously committed to training, on both a formal and informal level. It provided well over 1,000 hours of training per executive, locally and at its overseas locations. The value of this training beyond the confines of the corporation itself can be observed by the fact that the vast majority of Sears personnel who voluntarily left the company did so to take jobs in related areas where their training and experience would be of value. Thus, Sears has served as a type of practical training school in the art of commercial development, introducing the latest techniques in merchandising, product development, warehousing, credit, finance, and facility development through its formal training courses and through the extensive program of ongoing training.

**The Transparency Effect**

The operations of a mass merchandiser have a multiplier effect in an economy not only as a result of the impact of their own operations but because of the ease with which others can adopt the same ideas and procedures. Mass merchandising technology is not patented. Its success depends not on secrecy but on innovation and integration. Thus, for the most part, it is transparent; it can be adopted and adapted by others to serve their needs.

The impact of the entry into the retail market by Sears in both Peru and Mexico illustrates this transparency. New approaches to advertising, merchandise display, product guarantees, and site location were adopted after Sears introduced them. Credit was offered on a broader basis after Sears began to do so. In fact, in both countries competitors adopted large portions of the Sears credit system for use in their stores, either through hiring Sears-trained personnel or gaining access to Sears credit manuals.

The transparency effect functions to make an economy more dynamic through increasing competition and providing access to innovation. Innovations must be introduced that appear to have an advantage over existing approaches. Because of its access to mass merchandising technologies tested and proven in major world markets, Sears serves as an important mechanism for transferring such technologies. If those innovations appear to be successful in the local market, then they may be adopted and adapted by others, thereby enhancing their impact on the economy. Credit provides a good example. The pressure of competition acted to motivate competitors of Sears to provide credit to a larger number of clients, as Sears was doing. Thus, buying power was expanded not only by the credit Sears made available but by the increased credit made available by the other department stores as well. Likewise, it could be expected that as Sears continued to introduce innovations such as construction and operating techniques for reduced energy usage, they would be picked up and used by the competitors if they represented improved efficiency or greater consumer appeal.

Thus, one of the major impacts on the economic development of mass merchandising technology lies precisely in its transparency. Once introduced and proven successful, it can be and is quickly adopted by others in the market, thereby multiplying its impact. Clearly, a continuing impact depends on access by Sears, Roebuck's affiliates to innovations in new and better approaches to mass merchandising.

## Conclusion

Sears, Roebuck and Co. introduced the concept of mass merchandising into Mexico and Peru using a system of technologies transferred through training, documentation, and the use of Sears International personnel. This three-pronged approach has provided for both maintenance of the system and innovation. It has given the local corporations access to the knowledge and experience of a major international merchandising system. New ideas and innovations, tested out in U.S. markets, have been made available for introduction and adaptation locally. The economies of scale generated by the experience of a larger organization have been available. And the problem-solving capabilities of skilled executives have been on immediate call.

The introduction of a mass merchandising system into Mexico and Peru turned a commercial sector oriented toward the elite trade into one much more oriented to the middle class. It changed the perception of the market from a static to a dynamic one. It helped develop local manufacturers and new products, made credit more broadly available, and increased the efficiency of resource use. It was done by transferring and adapting the techniques and innovations that have worked well in the U.S. market to the Latin American market. This represents both a danger and an advantage. The danger, of course, is that the techniques will not work or that they will not be considered innovative enough to justify a foreign presence. The advantage is that the innovations, such as a data-processing system or solar heating panels, come as part of an overall system which reduces their individual cost and which includes constant training, access to technical assistance, problem solving, and new ideas.

That there have been and are problems cannot be denied. Local personnel at times have seen Sears International as overriding local ideas with a foreign system. Certainly, the type of assistance provided have had to change as the local corporation and economy develop, as illustrated by Mexico's generation of most of its own product ideas. There has been a desire on the part of some to reorient Sears toward the upper end of the market, whereas others seem to realize that a foreign corporation is most effective when it contributes directly to the achievement of national goals such as the expansion of the middle and lower-middle class, promotion of commerce in small towns, development of local manufacturers, and maintenance of dynamic competition in the local market.

Sears Mexico emerged from the doldrums of the 1960s and by 1978 appeared to have found a new direction for itself, one well suited to the contributions it has to make. Sears Peru was holding its own in terms of image and market, but the handwriting was on the wall. Its competitors, particularly for the middle and lower class markets, were expanding. It could not expect to maintain market position or staff morale over the long term without a dynamic growth-oriented program, as Sears Mexico learned. Yet it was hamstrung by a depressed economy, an unclear political situation, the geography-imposed costs and difficulties of expanding beyond Lima, and competition with larger, richer, and more stable markets for corporate investment funds.

From an economic development point of view, two factors are of primary importance. First is the concept of mass merchandising itself, with its commitment to making products available to the middle and lower-middle class and, by so doing, stimulating the growth of that class. The temptations to move toward the upper end of the market were obvious in Mexico, but there Sears recommitted itself to the mass market with the adoption of its expansion plans for the smaller urban areas of the country. In Peru, Sears took major steps in introducing the concept of mass marketing, but then faced the decision as to what its future should be. The question was should it continue to view itself as a mass marketer with the geographic and lower-middle class market expansion such a vision would entail.

The second factor is that the technologies transferred as part of a mass merchandising system are transparent. Not only is a monopoly position for foreign investors not possible, their presence and the type of technologies they transfer act to make the local market dynamic and competitive through the transparency effect.

The center of the mass merchandising philosophy is the consumer. In this it differs significantly from the traditional approaches to industrialization in the developing world, which have assumed an existing and rather static demand and concentrated efforts on production. By focusing on demand first, the mass merchandiser has sought out those areas in which it exists and can be expanded, and then directed the productive process to meeting consumer demand efficiently and effectively, which can make mass merchandising a consistently innovative force in economic development.

## APPENDIX: SEARS MASS MERCHANDISING TECHNOLOGIES

The technologies developed by Sears U.S. in over 90 years of operation are briefly described below in terms of their function within the system. They consist of six basic technologies: merchandise, operation, credit, facility development, personnel, and finance.

### Merchandise

The Merchandise System is responsible for the selection, purchase, and display of every item sold in a Sears store. It also encompasses consumer education about the item and is responsible for guaranteeing product quality and reliability. It consists of the following components.

*Merchandise research* provides the analytical information necessary for Sears buyers to decide what type of product or service is desired by the consumer. It also does basic site location studies.

*Merchandise comparison* conducts on-site studies to compare Sears products, physical facilities, and services with those of its competitors. Such areas as stock levels, price, quality, and selection in specific lines between Sears and its competitors are covered.

*Planned buying* incorporates the information from merchandise research and merchandise comparison on customer preferences and competitors' lines with analyses of previous sales, market forecasts, and wholesale product price information to provide the Sears buyers with the information necessary to make a decision on what color, style, price, and quantity of product to order and when to take delivery. Its primary function is to ensure accuracy in relating purchase of merchandise to projected sales so that inventory turnover is maximized.

*Product and source development* are among the key functions that separate the mass merchandiser from the retailer. Product development may take the form of an addition to an already existing line of products, a change in product characteristics, the addition of safety features, or the development of a Sears exclusive feature for a completely new product. To carry out these functions, buyers develop and maintain a close relationship with their product sources. The buyers' role is not simply to request a new or modified product, but to assist the manufacturer in developing a product that will meet Sears standards and specifications. To do this, the buyer may offer technical, managerial, and even financial assistance.

*Inventory control* in conjunction with planned buying is designed to maximize inventory turnover while maintaining adequate stock to supply customers.

*Sales promotion and advertising* is responsible for educating the consumer as to the type, selection, quality, feature, and prices of goods offered for sale.

*Visual merchandising* covers all aspects of the presentation of both products and services to the customer in the store as well as basic interior layout of the stores. It is estimated that no more than 20 percent of all customers are attracted to the store by its advertisements. Therefore, visual display of products and the proper design of the store for that display are key factors in sales.

*Purchasing management* is responsible for acquiring all of the fixtures, displays, and operating equipment required within the Sears stores, warehouses, and offices.

**Consumer credit**

All credit systems are very sensitive. They depend on the ability to establish the correct level of credit availability, identify good credit risks and educate them in the proper use of credit, service accounts rapidly, and evaluate account delinquency. There are six subsystems in the Sears credit system designed to fulfill these functions.

*New accounts* sets the procedures for issuing credit, conducting investigations of credit applicants, and deciding on their credit worthiness. This department also educates the consumer on the proper use of credit.

*Account servicing* is responsible for processing credit sales, billing and auditing, and maintenance of credit accounts. It depends on the maintenance of detailed and accurate records in order to ensure the steady flow of payments necessary for the operation of the system.

*Account collection* deals with delinquent accounts, including determination of the reasons for delinquency and establishment of satisfactory payment schedules. In addition, it provides the reviews of trends in account delinquency that are crucial for the development of effective credit policies.

*Income and expense management* establishes the procedures for the efficient operation of the Credit Department. It keeps a constant check on productivity through comparative review of payroll, personnel, and productivity goals, supplemented by productivity studies and standards.

*Personnel development and training* has the task of providing credit management with a systematic method for training all credit employees, including a credit management training program for credit executives.

*Consumer research* evaluates Sears credit policies and procedures based on general economic and market research as well as specific research into credit customers' knowledge of and attitudes toward Sears credit programs. It has been useful in alerting management to problems with credit service, which could then be corrected. It is also responsible for new system development, such as computerization of credit account handling.

## Operations System

Because it is primarily a support system for the selling function, Operations is often overlooked in the analysis of a mass merchandising distribution system. However, it plays a critical role, not only in ensuring that merchandise in good condition is delivered to the selling floor at the time desired, but in repairing and servicing the merchandise and, most importantly, maintaining a constant control over payroll and expenses. There are eight technology subsystems within Operations.

*Payroll and expenses* is the critical underpinning for the entire retail system, for only when constant attention is given to analyzing and controlling expenses, can the efficiency of the system be maintained.

Payroll management is defined as the effective utilization of the skills of the employees, based on an evaluation of the needs and requirements of the operation. For example, significant reduction in the number of employees and therefore the cost of the payroll could be achieved by reducing the amount of time spent in training. But doing so would reduce the efficiency and quality of service and products on which profits depend. Likewise, limiting expenditures in other areas to a minimum may not be the way to improve profits. A decision to decrease advertising in a slow period may further decrease sales and thereby actually increase rather than decrease costs as a percentage of sales. A decision to expand advertising prudently in such a period may increase sales and ease the company through a slow period.

Decisions such as these can be made only by persons with in-depth experience with and understanding of how the system operates and what the critical interrelationships are. Thus, payroll and expense management requires familiarity with all of the technology systems within a mass merchandising operation and constant attention to an evaluation of their costs and benefits.

*Traffic and transportation* is responsible for the physical distribution of goods from the source to the various warehouses and stores.

*Warehousing and materials handling* deals with all Sears inventory other than that in the stores themselves. Merchandise arriving from suppliers or manufacturers is checked for defects, inventoried, and delivered on demand to the stores or customers.

*Receiving, marketing, and shipping* is responsible for goods in the stores and handles receipt, marking, and shipping to customers.

*The Selling System Guide* is designed to give salespeople a detailed reference book on sales policies, procedures, and adjustments so that quick and accurate service can be given to the customer.

*Product service and repair* is integral to the Sears operation, for it guarantees customers that repair and servicing of their purchases are easily available from persons trained to do it.

*Safety and security* establishes the procedures necessary to guard against conditions in stores and other facilities that could cause customers accidents or injury. It also is responsible for protecting Sears assets against losses from fire, natural disaster, fraud, and theft.

*Data processing* provides the various functional departments of Sears with modern computer services. The installation of data-processing facilities enabled Sears U.S. to double its sales without the increases in personnel and administrative procedures that generally follow such sales increases.

## Personnel System

The Personnel System is responsible for the selection and training of personnel, placement, career patterns, compensation, benefits, and morale. There are six basic technology subsystems that make up the Personnel System:

*Testing, selection, and placement* concerns the testing of both temperament and mental abilities of nonmanagement personnel. Through its years of experience, Sears has developed two basic tests. They have been adapted for use in Latin America. Basic standards have been established for each job classification in order to try to ensure that a person's abilities and temperament match the job.

*Compensation administration* has as an aim the maintenance of a system of compensation that is perceived as fair and viewed as paying above the average for comparable work. This is considered the key to maintaining high quality personnel and good morale.

A compensation manual for all support personnel outlines the basic relationship, in terms of pay and benefits, which should exist between various jobs. Actual wages are determined by areawide wage surveys done to determine the range of pay that persons in similar jobs with similar skills are receiving. The compensation system for executives is based on volume of sales, store size, and factors relevant to their particular jobs. Sears also has a profit-sharing plan to which both the employer and employee contribute.

*Policy and benefits* integrates the standard Sears policy of personnel benefits with the legal requirements in the area of operations.

*Employee relations* deals with those factors affecting employee morale. Individual review sessions are held on a yearly basis.

On the organizational level, questionnaires on employee morale are periodically circulated, in which employees are asked to comment on the supervision they have received, their feelings about their job, the amount of work, working conditions, strengths and weaknesses of the Sears system, and so forth. The questionnaires are analyzed and a report to employees is prepared and disseminated. It includes not only a summary of the responses to the questionnaire, but also what the company proposes to do about the problems that have been identified. The questionnaire is valuable in helping the company to spot problems before they become serious. Reporting the results to the employee puts pressure on the company to take action to deal with the problems.

*Technical training* provides the training programs necessary to equip the average person to perform most jobs in the Sears system. It also provides for the training of technicians. Included in it are the basic courses that all employees involved in store operation receive as well as all of the courses offered by the SEI.

*Management development* has the responsibility of providing for the development of management personnel. Two basic programs exist. One is the year-long Management Training Program for recent university graduates. The second program is designed for persons already within the Sears system who have demonstrated ability and interest. Both lead to participation in the reserve group from which new executive personnel are drawn. Special attention is given to developing a career pattern for reserve group members, which gives them experience in all areas of the system but also leads to concentration in the area of the employees' particular abilities and interests.

**Financial System**

Finance is responsible for providing the data and analysis necessary to evaluate the company's financial position, needs, and growth possibilities and for borrowing the money required by the corporation. Finance consists of four technology subsystems.

*Accounting procedures* is responsible for devising uniform methods of recording, classifying, summarizing, and reporting basic financial information.

*Internal auditing* is the mechanism for testing compliance with Sears policies and safeguarding all assets of the corporation.

*Budgetary procedures* are developed to aid managers at all levels to plan their purchases and their payroll over a six-month period.

*Cash management* is responsible for ensuring that cash is available for existing operations as well as for financing expansion programs.

**Facility Development System**

The Facility Development System provides for the development and modernization of all Sears facilities, including stores, shopping centers, warehouses, and offices. It may also take responsibility for the operation and maintenance of these facilities, particularly shopping centers. There are four technology subsystems within Facility Development.

*Locational analysis* is the first in the remodeling or building of a store. It is done in conjunction with site location studies of market research. Before a decision can be reached on the advisability of such a step, analysis must be made of the population and income levels in the area, their projected growth rates, site availability and cost, transportation, competition, and possible overlapping with the sales areas of existing Sears stores.

*Feasibility studies* are conducted before a final decision is made. Soil tests are carried out to determine design variations for the building. Investigations of legal requirements, existing and projected tax rates, surveys of hot lines, easements, and utility lines, and traffic analyses are made, including an income and expense statement, cash flow, and mortgage analyses. If the project is a shopping center, initial leasing efforts with major space users will also be carried out. The final decision depends on financial projections, initial building design, and review of legal and tax situation.

*Design and construction management* is responsible for store design, which must be done with one primary purpose in mind—the efficient sale and servicing of merchandise. Sears has developed certain principles for the basic design and the interior layout of its stores: Stock rooms must be placed within easy access to both loading platforms and selling areas; credit offices are placed close to areas that do the most sales on credit; parking is laid out for quick and easy access to both the store and connecting roads. In a warehouse, design specifications are based on knowledge of the size and type of equipment and merchandise that will be used in it.

Actual construction may be contracted out, but its management is the responsibility of Sears, including overseeing actual building, ensuring financing, controlling costs, and guaranteeing that all municipal codes and regulations are met.

*Operations administration* varies according to the project involved. In most cases, for a store warehouse, it will consist of training the building maintenance workers. It will also include periodic checks on building safety and protective equipment and procedures. For a shopping center, it may include the actual administration of the center, covering tenant management as well as preventive maintenance.

## NOTES

1. This study was researched and written in 1978. Therefore, the past tense is used in referring to the 1978 period.

2. William Glade and Jon G. Udell, "The Marketing Behavior of Peruvian Firms: Obstacles and Contributions to Economic Development," in *Markets and Marketing in Developing Economies,* ed. Reed Mayer and Stanley C. Hollander (Homewood: Richard D. Irving, 1968), pp. 158–60.

3. Peter F. Drucker, *Management* (New York: Harper & Row, 1973), p. 56.

4. Ibid, pp. 53–54.

5. Ibid, p. 54.

6. Gordon L. Weil, *Sears, Roebuck, U.S.A., The Great American Catalogue Store and How it Grew* (New York: Stein and Day, 1977), p. 169.

7. Ibid, p. 180.

8. *Business Latin America,* Dec. 27, 1978, pp. 412–14.

9. Ibid, pp. 411–15.

10. "Mexico Annual Supplement," *Economic Review of Mexico,* 1977, p. 6.

11. Daniel James, "Sears, Roebuck's Mexican Revolution," *Harpers Magazine* June 1959: 2.

12. Unless otherwise noted, statistics on Sears operations have come directly from, or been calculated by the author on the basis of, figures provided by Sears Mexico, Sears Peru, or Sears International.

13. Richardson Wood and Virginia Kayser, *Sears, Roebuck de Mexico, S.A.* (First Case Study in an NPA Series on U.S. Business Performance Abroad, National Planning Association, 1953), p. 11.

14. William R. Fritsch *Progress and Profits, The Sears, Roebuck Story in Peru* (Washington, D.C.: Action Committee for International Development, 1962), p. 13.

15. Ibid, p. 31.

16. Ibid, p. 30.

17. Ibid, p. 10.

18. Douglas F. Lamond, "Mexico" in *Public Policy Toward Retailing, an International Symposium,* eds. J.J. Boddewyn and Stanley C. Hollander (Lexington: Lexington Books, 1972), p. 233.

19. Compania Privada de Investigacion de Mercados, Peruana de Opinion Publica.

20. *Guidelines for the Study of the Transfer of Technology to Developing Countries* (UNCTAD United Nations, New York, 1972, p. 5) provides a good definition of technology: (1) Capital goods, including machinery, and productive systems. (2) Human labor, usually skilled manpower and management, specialized scientists. (3) Information, of both a technical and commercial character, including that which is readily available, and that subject to proprietary rights and restrictions.

21. "Industrias Reunidas Comenta sobre la Contribucion de Sears a la Industria de Artifactos Domesticos en el Peru," Lima, February 16, 1965.

22. Eduardo Romero Accinelli, "Hermanas en Los Negocios," Lima, June 7, 1976.

23. "Industrias Reunidas."

24. *Guia de los Mercados de Mexico,* Novena Edicion, 1976–1977, Marynka Olizar, Mexico, D/F/ fr. cuadro IV, Guia de los Mercados, p. 28.

25. James, "Mexican Revolution." *Harpers Magazine*, June 1959, p. 2 (reprint).

26. Paul McCracken, James C.T. Mas, and Cedric Fricke, *Consumer Installment Credit and Public Policy, Michigan Business Studies, Vol. XVII, No. 1* (Ann Arbor: University of Michigan, 1965), pp. 27–28.

27. The population figures used are for Mexico and for the Lima metropolitan area.

28. 1953, six years after the founding of Sears Mexico, was selected in order to avoid the effect of the rapidly increasing number of credit accounts that occurs during the first year of a store's existence.

29. The figures used in this section are from the *United Nations Statistical Yearbook, 1961 and 1970,* the Office of Population Statistics, Interamerican Development Bank, and the Office of the United Nations, *Demographic Yearbook.*

30. McCracken, "Consumer Installment Credit," p. 3.

31. The methodology used to calculate this figure is, of necessity, very rough. The purpose of the exercise is to show an order of magnitude and to begin to develop a methodology. The first step is to obtain information for the following:

$A$ = Lines of merchandise

$B$ = Major sources (account for more than 20 percent of products sold in line)

$C$ = Percentage of total production of major sources bought by Sears (by production is meant all items produced in the factory, not just those sold to Sears)

$D$ = Number of persons employed by major sources

$E$ = Percentage of sales of merchandise made on credit

$C \times D = F$ = Number of employees producing products Sears buys. (The author recognizes that this overlooks the fact that capital/labor ratios in the production process vary from product to product. Dealing with this issue is a refinement that should be added when possible. However, the author assumes for purposes here that over a wide list of products, the differences from product to product should average out.

$F \times E = G$ = Number of jobs attributable to credit availability.

$0.7 \times G = H$ = Sears estimate of the amount of sales that would be lost were credit not available is 0.7. Number of jobs dependent on credit.

$G - H = I$ = Number of jobs attributable to credit that would continue to exist.

$F - G = J =$ Number of jobs attributable to cash payment.

| | |
|---|---|
| 1,388 | Jobs directly attributable to availability of credit by Sears for appliances, home furnishing, and improvements (*G*) in Mexico. |
| 0.70 | Percentage of sales estimated to be lost were credit not available. |
| 972 | Jobs attributable to the availability of credit (*H*). |

32. "Market Research Report," Sears, Roebuck del Peru, Lima, August, 1978.

CHAPTER 4

# Technology Transfer by the Bechtel Organization

*John C. Stephenson*

## INTRODUCTION

This study of the Bechtel organization examines the wide range of engineering and management services provided by that organization; reviews the technology transfers that take place in the course of Bechtel's engineering, construction, and management of major projects in other countries; gives examples of its activities; and summarizes the views of Bechtel's transfer activities by a number of their clients in one host country.

Interviews with key Bechtel staff in seven major operating groups of the company located in San Francisco and Los Angeles, California, and in Houston, Texas, and meetings with four Bechtel clients in Mexico provided the basic information for this study. The observations and comments are the author's responsibility.

Bechtel designs and builds large scale industrial plants, utilities, and other infrastructure projects that contribute to a nation's industrial and economic development. In the course of project design, engineering, and the actual construction, significant technologies are transferred, ranging from basic construction craft skills to sophisticated engineering and management. In its advanced technology transfer activities, Bechtel infuses new knowledge into another organization to enable that organization independently to design, construct, and manage highly complex projects. In effect, it transfers its own skills to clients so that, in time, they can carry out the very type of work that Bechtel originally was hired to do.

The essential nature of Bechtel's technology is its ability to manage complex technologies and large facilities of many kinds to produce consistent results of high quality and to do this on time and within budget.

The scope and type of technology transfer in the course of any specific project depend on several interrelated factors: the technology transfer objectives and plans of the client; the education and experience level of the recipient; the nature of the project and Bechtel's role in project execution; the duration of the project; and the complexity of the technology involved. The long-term utility of the transfer experience for the client depends on the attitude and motivation of the recipient during the learning process and the opportunity of the recipient to utilize on a continuing basis the knowledge gained via technology transfer.

Bechtel has found that the most effective way to transfer knowledge and build ability in individuals is through on-the-job training supplemented by specifically designed instructional programs. To accomplish advanced technology transfer, integrated Bechtel-client project teams are used to carry out work assignments. To maximize transfer in these integrated teams, a one-on-one counterpart system—national participant and Bechtel professional—is used. The goal of the integrated team is to transfer gradually the leadership role in the project from Bechtel to the client as Bechtel moves to an advisory role. In thermal power technology transfer, for example, Bechtel has found the following elements vital for success in making the transfer: the use of an approach to project management that involves to the maximum extent existing host nation technical and management resources; carefully planned and formalized program for technology transfer; and multiproject involvement of national professionals in project engineering and management of construction over an extended period of time to allow for the planned progression and transfer of responsibility.

It is difficult to define and to measure success clearly in technology transfer. One measure is the number of persons trained. Some tens of thousands of foreign nationals are estimated to have been trained in the course of Bechtel work overseas during the last 35 years. Over the last ten years alone, the advanced technology transfer operations of Bechtel's thermal power group have resulted in the on-the-job training of over 5,000 national technical and professional personnel.

Disciplined work habits, basic construction craft skills, planning skills, and practical management experience are among the by-products of most Bechtel assignments. Although the individuals with these skills may subsequently be widely dispersed within a country once the specific project work is completed, they constitute an important addition to the trained resource base of those countries. Advanced technology transfer programs enhance the development of technological self-sufficiency and sophisticated professional skills. From the point of view of the client and the host nation government, the transfer of service technology by Bechtel is a positive contribution to their organizational ability and to technological and national development.

Agreements to provide such services are entered into by means of arms-length negotiations between independent entities. Like most consulting activities, projects are of limited duration, and contracts entered into clearly define what is to be provided, the terms under which it is to be provided, and at what cost.

Transfer of service technology is essentially a person-to-person transfer. It is not done through licensing agreements, manuals, drawings and specifications, or computer systems alone. The personal "chemistry" between the transferor and the recipient is most important. Those transferring the technology require a supportive family environment, a "people" orientation in addition to high technical professionalism, and a sensitivity to the culture and environment into which the new technology is being introduced. Introduction of new management methods and techniques can pose problems of culture shock for both parties in many countries. In addition to technical proficiency, successful recipients of advanced technology training require adaptability and willingness to accept the new, and need to gain the wisdom to acknowledge what they know and what they do not know. Finally, in both the transferor and recipient company or institution, top management must understand the magnitude of the technology transfer effort and its side-effects, and be prepared to support it vigorously by word and deed if the transfer is to succeed.

## OVERVIEW OF THE BECHTEL ORGANIZATION

The Bechtel group of companies is one of the largest engineering and construction enterprises in the world. It began as a family-owned railroad construction business in 1898. Over the years the scope and variety of its work increased. By the 1930s Bechtel was taking part in such construction milestones as the Hoover Dam and the San Francisco-Oakland Bay Bridge. By 1940 Bechtel had, in addition to construction, broadened its area of activity to include engineering and procurement. Today, it offers these and a number of other related services.

Bechtel's first international project came in 1940 when it began to work on a pipeline system in Venezuela with joint venture partners. This was followed by a refinery in Bahrain and other projects for refineries, harbors, and power plants in the Middle East.

Bechtel today is an international organization with offices, subsidiaries, and affiliates in 20 countries around the world. It has carried out projects in some hundred nations on all seven continents. Revenues in 1979 were $6.8 billion, of which just over half came from outside the United States. During that year 116 major projects were underway in 21 countries. In 1979 there were approximately 35,000 permanent employees on the Bechtel nonmanual (noncraft labor) payroll; about 7,000 of these permanent employees came from over 100 different countries outside the United States. If hourly paid workers at Bechtel job sites worldwide are included in the total, peak employment in 1979 was approximately 105,000.

The Bechtel organization includes three major operating units, the largest of which—the Bechtel Power Corporation, with four divisions—accounts for about half of Bechtel's employees and revenue. The other major operating units include Bechtel Petroleum Inc., which has three divisions, and Bechtel Civil and Minerals, Inc., which has three operating entities. Research and Engineering is another significant operation with the group. A fourth major company, Bechtel Investments, Inc., handles investment activities. A simplified structural representation of the Bechtel Group is shown in Figure 4.1.

The four operating divisions within the Bechtel Power Corporation have provided engineering, procurement, and construction (EPC) services for many nuclear, fossil fuel, and geothermal power-generating plants around the world and for some electrical transmission and distribution facilities.

Nuclear Fuel Operations provides EPC services in all areas of the nuclear fuel cycle, from enrichment of uranium to the decontamination and decommissioning of facilities, together with the long-term safe storage or disposal of radioactive wastes.

The Refinery and Chemical Division and its related Petroleum and Chemical-Eastern Division together have worldwide responsibility to design and construct petroleum refineries and petrochemical plants; forest products-, pharmaceutical-, fiber-, and rubber-processing plants; food and beverage facilities; liquefied and synthetic natural gas plants; energy-related tar, sand, and oil shale extraction projects; facilities for gasification and liquefaction of coal; and industrial plants and facilities for manufacturing transportation equipment.

The Pipeline and Production Facilities Division encompasses oil and gas field development from wellheads through field processing; marine work, including production and drilling platforms; and long-distance transportation systems for oil, gas, petroleum products, coal, and minerals, including slurry pipelines.

**Figure 4.1. The Bechtel Organization.**

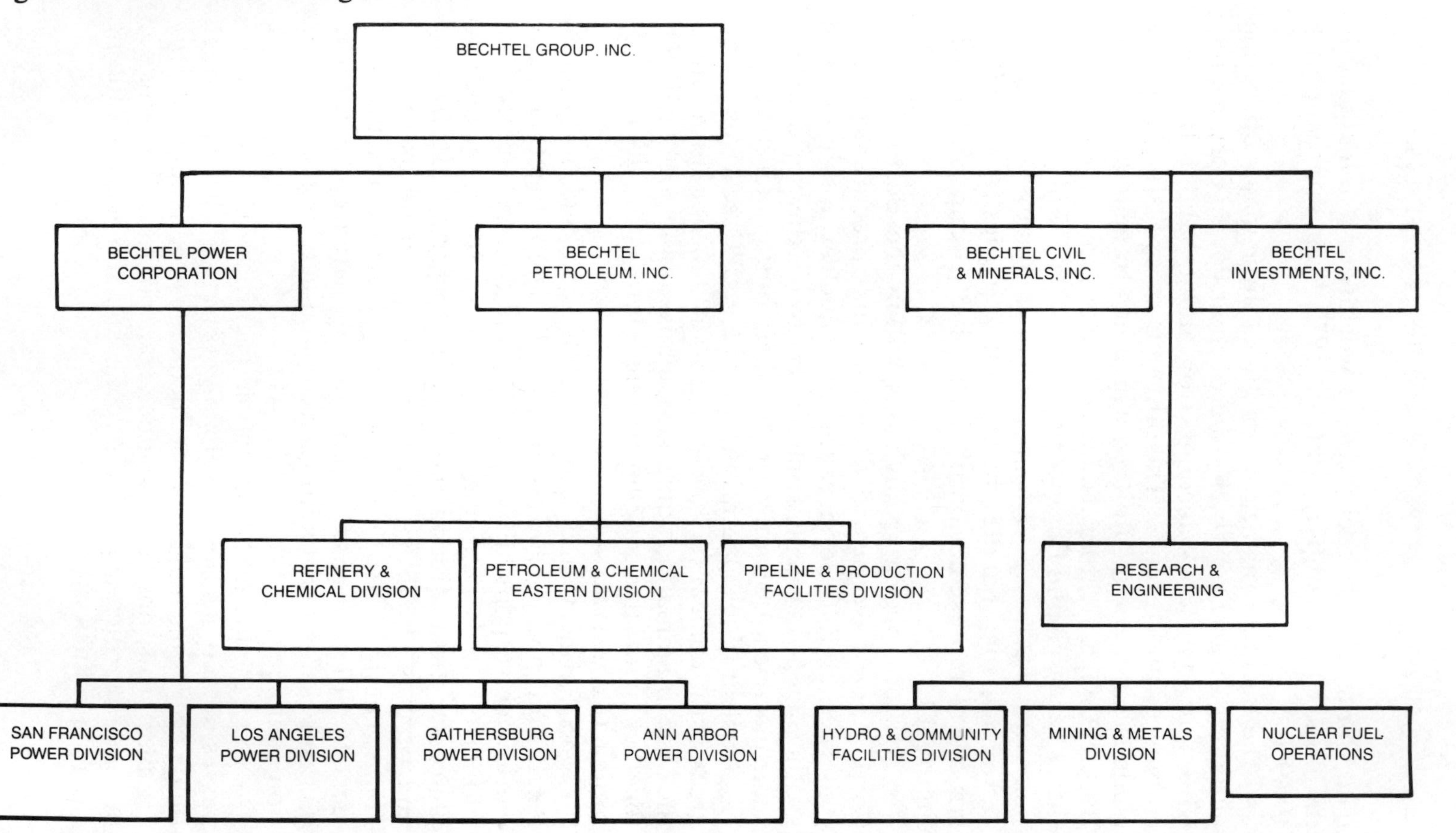

The Hydro and Community Facilities Division provides EPC services for hydroelectric power development; water conveyances; rapid transit systems; railroad, airport, port, and harbor development; telecommunications; solid waste treatment; commercial and industrial buildings; and hotels and hospitals, residential housing, and other types of infrastructure projects.

The Mining and Metals Division provides EPC services for facilities in processing metallic and industrial minerals from mining through beneficiation and extraction metallurgy; metal producing, refining, and shaping; and the manufacture of stone, cement, and clay products.

Finally, Research and Engineering maintains and further develops Bechtel's advanced engineering and process technology, and, in addition, provides professional services in materials testing, environmental engineering, and regional development.

Important support services involved in technology transfer activities are Procurement and Finance. Procurement works with equipment and material suppliers worldwide, and seeks the best pricing arrangements, on-time delivery, and high quality products and services for its clients and projects. Bechtel's financing specialists help its clients and its own project personnel develop the financing plans and securities structures necessary to attract favorable funding commitments to meet a client's financing need.

## TECHNOLOGIES INVOLVED IN ENGINEERING AND CONSTRUCTION

As the brief description of activities suggests, the technologies employed by Bechtel span a broad range of professional engineering, scientific, and management disciplines. The professionalism of Bechtel's staff is seen in the fact that in 1979, one-third of its 35,000 permanent employees were college graduates, over one-fourth of the total were engineers, and one-eighth of the employees had one or more advanced degrees.

Technology in terms of the Bechtel organization, however, is basically the "know-how" that permits the application of scientific findings to the creation of a practical facility or the performance of a specific task through planning, engineering, construction, and management. This is the essential nature of Bechtel's technology—the ability to manage complex technical requirements to produce results of high quality and reliability on time and within budget.

The building of a nuclear power plant is one example of a complex Bechtel technical project. The time required from preproject studies through engineering design to construction and actual start-up of such a power plant can be from 8 to 12 years, depending on where the plant is located and on the regulatory situation. From the engineering point of view, the project represents a several million-person-hour engineering program, with a peak design office team of 200 to 300 people preparing over 5,000 drawings and 400 specifications—and in the course of 5 years, handling over 30,000 vendor equipment drawings. The project is managed by a project manager who is supported not only by engineering and procurement personnel but also by a team of several hundred professional and technical personnel in the field office. The construction work force itself numbers in the thousands. Typically, several hundred equipment specifications and anywhere from 6 to 12 separate construction and installation contracts are involved in a Bechtel project. The cost of a 900-megawatt pressurized water nuclear generating unit projected in 1980 dollars is on the order of $1 billion. Escalation of costs and interest on money borrowed while the plant is under construction account for more than 40 percent of that total. Therefore, the pressure is heavy on project management to bring the nuclear power station into service in the shortest possible time.

The ability to manage such a project on time and within budget is a technological art. This is especially true on an international project where, in addition to a Bechtel design office in the United States, there is a project design office in the host nation handling some portions of the work, plus another design office at the construction site handling other design details. To add to the complexity, there can be hundreds of suppliers throughout the world contracted to provide project equipment and materials to specifications and on schedule, and contracts with many independent construction and installation companies. Management skills and project control systems of a high order are required to ensure the necessary planning, communication, coordination, and control for a successful project.

Bechtel's know-how, acquired over many years and in many countries, ranges from superior construction techniques, such as shortcuts in preparing concrete forms, to complex computer programs for solving engineering problems or scheduling multiple interrelated activities. This management knowledge is the product for which Bechtel is basically hired—that is, the ability to execute complex projects successfully.

It is not possible to categorize the diverse technologies employed by the Bechtel group simply. For example, some 150 professional skills and levels are involved in the design and construction of a typical nuclear power plant. A clearer understanding of the nature of Bechtel's work can be gained by reviewing the basic steps involved in a project. A comprehensive assignment begins with basic project feasibility analysis, continues through construction and staffing of the plant, and ends with a successful operating installation. Bechtel activities can include:

- *Preproject studies:* Market studies, preliminary technical-economic evaluations, natural resource surveys, human resource evaluations, country sectorial analyses and interindustry studies
- *Feasibility analysis:* Detailed technical market evaluations, laboratory or semiindustrial process testing, regional and/or location studies, economic and financial project evaluations
- *Process or basic engineering design:* Engineering reconnaissance and investigation, project process and plant site selection, preliminary design and layout, time schedules, cost estimates
- *Project planning:* Master planning, financing, project management and organization, procurement, personnel training, project control systems, start-up manuals.
- *Detailed design engineering:* Design models, construction drawings, specifications for materials, preparation of bids for contracts and subcontracts, bills of materials, and material requisitions

- *Procurement*: Development of a qualified bidder's list, bid solicitation, and evaluation; purchase order and subcontract issuance and administration; expediting and surveillance; receiving, inspecting, and warehousing
- *Project control:* Establishment of flows of data and information in comparison with the project plan to detect deviations and allow early correction before serious consequences develop
- *Construction*: Recruit, hire, and train as required a labor force to carry out the actual construction work under Bechtel supervision and responsibility when acting as a prime contractor
- *Construction management:* Managing and administering contracts/subcontracts in the field, preparing construction budgets and schedules, staffing and organization for construction, monitoring performance
- *Start-up services:* Developing a test program plan and schedule in close liaison with engineering and construction, administrative and functional tests, system verification tests, operation tests, deficiency resolution
- *Preparation for facility operations and maintenance:* Training of plant operators, supervisors, plant engineers, and plant maintenance personnel; planning spare parts, fuel, and other consumable materials; ensuring that plant manuals, "as built" drawings, and plant procedures properly reflect safe and efficient operating practices
- *Operating management services:* Where appropriate, operating the plant for a limited period of time; trouble shooting or operations consulting, which can involve modifying the plant systems, changing plant operations to make them more efficient, and developing new procedures and operating manuals.

These activities are described later in examples of Bechtel construction projects involving technology transfer.

## TECHNOLOGY TRANSFER

Technology transfer can be defined differently in different settings. Some define the term as the transfer of physical objects and installations; others include the introduction of disciplined work habits and practices to indigenous populations in a less developed society. This can occur during the course of the construction of a major natural resource facility in a primitive area; for example, the building of a hydroelectric power plant, oil, or gas production installation or mine. Others consider that technology transfer begins when some kind of formalized training takes place and explicit skills, such as carpentry, welding, form making, or electrical skills, are taught. Finally, there are those who consider that "real" technology transfer takes place only after developing country nationals complete prescribed levels of education and training and the trained professionals then begin to apply their knowledge in practical engineering work with the help, guidance, and supervision of experienced practitioners. Bechtel's technology transfer activities span this broad spectrum.

**Background**

Some form of technology or know-how transfer is inherent in all facets of engineering and construction business. Local labor and skilled workers customarily are hired at the job site where possible. There, on-the-job or more formalized training is given as required to get the job done. This was as true at the beginning of the Bechtel organization in 1898, when it was involved in railroad construction work in the western United States, as it is today, when almost 80 percent of total employment on Bechtel projects represents those of associated local contractors or temporary on-site employees. The scope of Bechtel's technology transfer activities has evolved over the years, determined largely by the requirements of specific assignments and the absence of needed resources to get the job done.

Formalized training activity within Bechtel probably dates back to its rapid development of the skilled work force needed for two large shipyards that Bechtel started up and operated in California in the early 1940s. Since that time, use of formalized training and the development of training programs by Bechtel have grown both in the United States and internationally, particularly in remote locations and areas of high unemployment.

Internationally, during the construction of the Trans-Arabian pipeline in the early 1950s and other Mideast projects, emphasis was placed on training workers from the local areas to provide support for Bechtel specialists. Although work skills training was limited to specific activities like pipe welding and installing pipe insulation or cable, some of the more promising workers were trained as instructors or as supervisors of small groups of locals. As Bechtel's overseas commitments and capabilities grew, the need for local training increased.

Most of Bechtel's early experience in training was concerned with construction skills. A natural expansion from that activity was the development of training programs for plant operations and maintenance. Programs in facility construction as well as operation and maintenance currently are underway in Algeria, Saudi Arabia, and Venezuela. Recent work in Indonesia and Saudi Arabia included training in administrative and clerical skills, computer operations, and skills required for the operation of community services. It is estimated that tens of thousands of foreign nationals have been trained on Bechtel projects in the last 35 years.

The more advanced forms of engineering and construction technology transfer as defined here have come about at Bechtel as an evolutionary development over the last three decades. During the 1950s and 1960s, most international projects were performed as "turnkey" jobs: Most of the engineering was done outside the host country. Technology transfer was then usually limited to on-site training activities for basic construction crafts, plant operations, and maintenance office staff skills.

In the 1960s and the 1970s, as host nations' technological capabilities expanded and awareness grew of the need to develop more fully their own national resources and manufacturing facilities, developing nations sought other ways to implement large scale projects that would draw upon those resources. They required overseas contractors to transfer ever more advanced technology in the course of accomplishing contracted work. Thus began the planned transfer of engineering, construction, and management technology to local organizations.

**Training**

In addition to on-the-job training for engineering and construction projects, Bechtel often establishes training schools at the project site. Formalized programs provide work/study experience for various skill levels of artisans, plant operators, technicians, and supervisors.

The steps that Bechtel follows in planning industrial/vocational training programs are outlined in Figure 4.2. The first nine steps shown lead to an assessment of the training need after an evaluation of project requirements and client preferences and a review of the available supply of workers—both national and foreign. The training plan that is developed (Step 10) reflects consideration of such factors as education and experience levels of the available workers, governmental requirements and guidelines, and cultural and environmental factors. Training planning includes the preparation of job descriptions, development of course outlines, selection of teaching methods, and design of methods for evaluating the programs.

In addition to a central Personnel Training and Development Group within the Personnel Department, many Bechtel divisions have their own employee-training staffs. From time to time subcontractors will be used to support Bechtel training personnel. In addition to courses in new employee orientation, construction safety, and supervisory training, Bechtel has developed a broad spectrum of programs for construction technicians and trades.

Operations and maintenance (O & M) training is another formalized training activity at Bechtel. It has been an outgrowth of the start-up and facility operations services provided by the company. Initial start-up and staffing of large grass roots facilities generally present great difficulties, especially in remote areas. Operating companies often need to rely heavily on contract foreign employees to fill O & M positions. The objective of Bechtel's O & M training is to prepare national employees to replace foreign employees in the shortest possible time.

Although the general development of training programs for operations and maintenance is similar to that outlined in Figure 4.2, specific elements of these programs include: early

**Figure 4.2. Planning and Implementation Steps for Industrial/Vocational Training.**

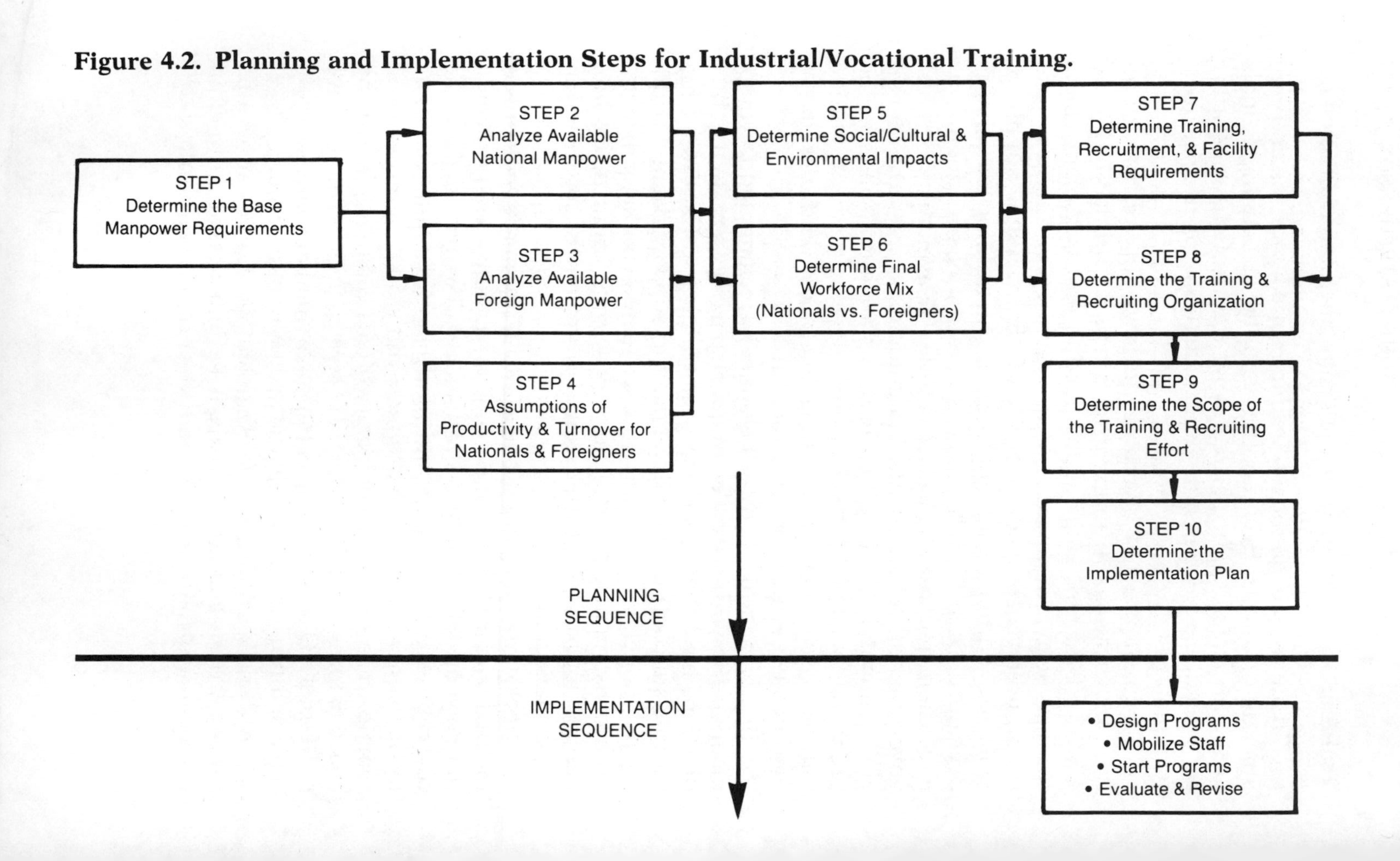

in-depth operations start-up planning in conjunction with project construction planning; an early start of training activities for national employees; development of internal plant policies, systems, and procedures to control the operations and maintenance of these facilities; continuous monitoring of training plans and schedules as the project progresses; and integration of the trained national work force with the Bechtel staff in the commissioning activities for the project.

At times, the construction craft training and O & M training are integrated into one program using common facilities and, to a limited extent, the same training personnel. Outstanding trainees from the construction work force can be subsequently trained as O & M personnel.

The training program development and implementation work described thus far relates to individual projects. Bechtel has applied the same approach to manpower planning for massive development projects involving large industrial installation complexes. For example, for a planned new industrial complex in Algeria, Bechtel completed a manpower development plan for the complex as part of the master project planning effort. The plan is a guide to the establishment of programs necessary to train the 26,000 operators who will be needed to run the proposed new factories and industrial installations, infrastructure, and new industrial city services.

In Saudi Arabia, Bechtel is responsible for the design of several facilities to provide managerial, administrative, craft, and operations training to Saudis who will be employed in the massive new industrial complex now being built at Jubail in the eastern part of the country. According to current plans, by the end of this century Jubail will have 16 primary industries and a community with a population projected to reach 370,000. Engineering, construction, management, supervisory, and equipment operator training is presently underway.

Figure 4.3 gives an idea of the magnitude of the training task involved in such large scale industrial development programs. The formalized training sequences that can be involved for six different skill levels of entrant employees are shown. The feedback of newly trained instructors from the program to various levels of instruction is shown with dotted

## Figure 4.3. Comprehensive Industrial/Vocational Training Program.

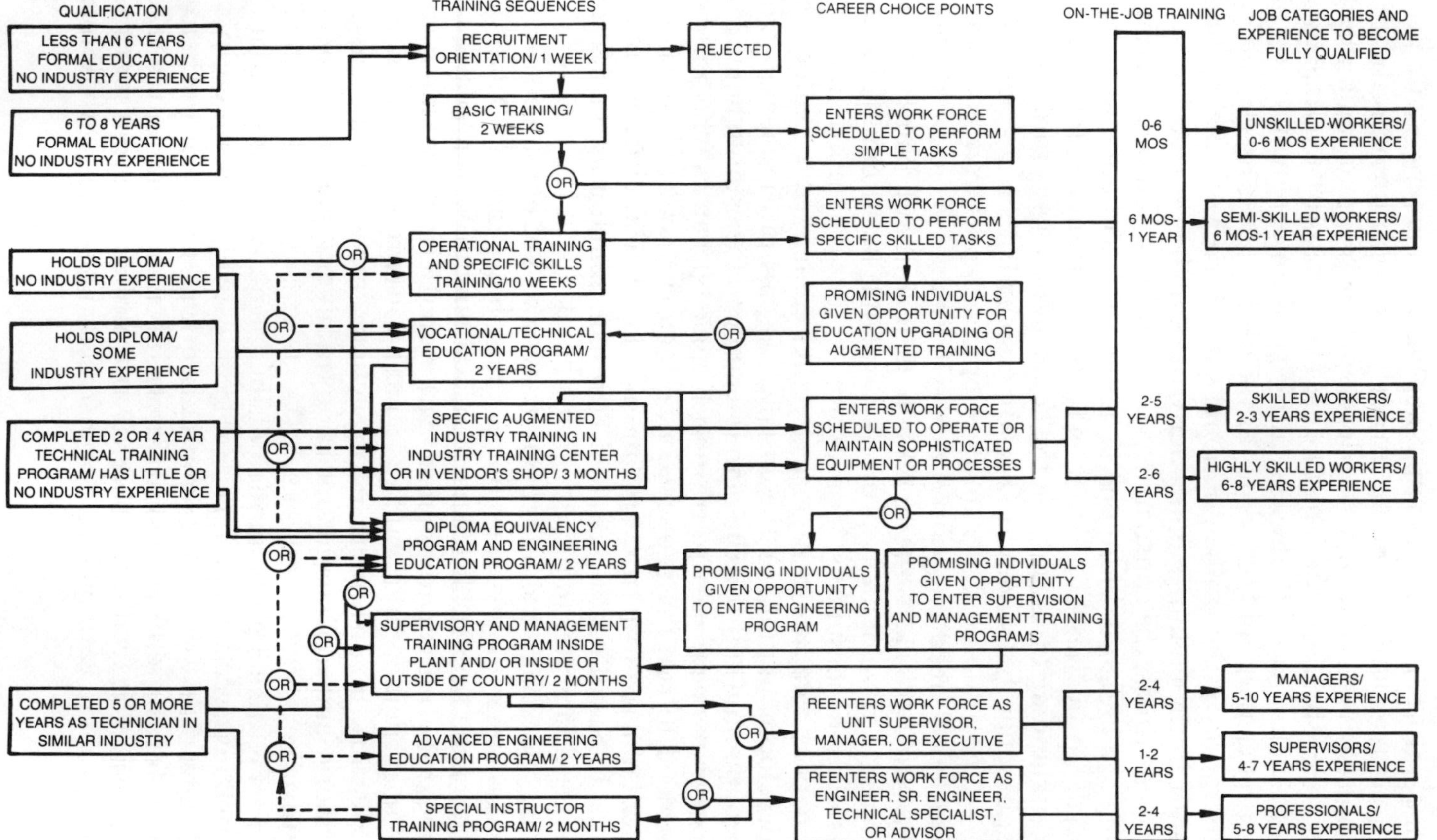


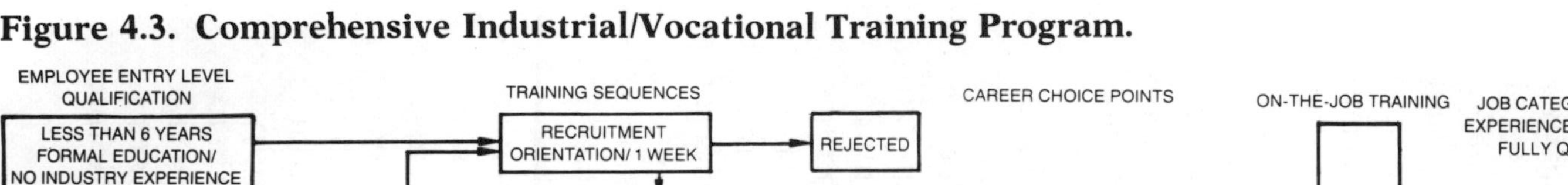

lines. At the completion of each formalized training sequence, there are career choice points for each employee. These decisions can lead to educational upgrading, additional formal training, or on-the-job training. The end-products of this overall program as shown are seven levels of job categories together with the qualifying experience required for them.

### *Examples of Training Experience*

For most Bechtel divisions, technology transfer has heavily concentrated on the development of various skill levels of skilled workers, technicians, plant operators, and supervisors. In addition, grass roots construction projects in remote areas usually require Bechtel to develop comprehensive basic supporting infrastructures, including public health systems and community facilities.

*Indonesian Nickel Project.* The recently completed Soroako Nickel Project of P. T. INCO in Indonesia is an example of a project carried out in a remote, underdeveloped area of the world where, in addition to engineering, construction, and severe logistic problems, there was the need to recruit and train great numbers of local manual and nonmanual staff. This large scale nickel mine and processing plant on the Island of Sulawesi has made Indonesia one of the major producers of nickel in the world.

The project location presented the challenges of mountainous terrain, dense tropical forest, and, at the plant site, sparse settlement and little preexisting infrastructure. In this project, the Mining and Metals Division had overall project responsibility for the entire production complex and permanent community facilities: process plant dam, hydroelectric power station, and residential areas. The new town constructed for the permanent plant operations staff included housing, communications systems, schools, and churches, plus commercial, medical, and recreational facilities. In addition, the division was responsible for the temporary on-site support facilities and services for 5,000 employees of the total construction force of 8,800.

An integral part of the commitment made to the government of the Republic of Indonesia by the client and by Bechtel was to provide formal and intensive training to a sizable percentage of those Indonesians who were or would be working as Soroako project employees. For its part, Bechtel established its own independent Training Department to join the client in training Indonesians for the project. Over the two years of formalized on-the-job training, more than 2,500 Indonesians were trained in construction crafts and the administrative techniques needed to complete the construction work. The trainee population peaked at about 1,000 when 20 courses were being conducted at 12 training areas.

Most of the trainees were unskilled or semi-skilled, or in any event had little or no experience in the skills they were learning. As the most critical, on-site job vacancies for skilled workers began to be filled by trainee graduates, increasing emphasis was placed on on-the-job training. At first, most of the formal training programs were conducted by Bechtel. Thereafter, by training exceptional Indonesian graduates of the skilled craft courses to become fulltime and assistant instructors, a rapid expansion in the scope and size of the program was made possible.

*Indonesian Liquefied Natural Gas Facilities.* In Indonesia, the Bechtel Refinery and Chemical Division had the responsibility for the design, procurement, and construction of two large liquefied natural gas (LNG) facilities at Arun and Bontag Bay. At Arun this involved the coordination of the efforts of some 8,000 persons, including 240 Bechtel employees and 2,500 subcontractor employees.

These Indonesian grass roots projects required one of the largest training efforts in Bechtel's history. Five training programs were established in different parts of Indonesia to train, certify to international standards, and send to those two projects more than 2,000 skilled workers: welders, pipefitters, electricians, boilermakers, millwrights, and instrument fitters. In addition, on-the-job training programs were developed at Arun and Badak to train hundreds of workers as carpenters, cement finishers, iron workers, equipment operators, and painters. After six months of training, people who had never used a tool more complicated than a machete were operating 350-ton cranes. These training programs paralleled construction schedules for each craft. A curriculum group was organized at Bechtel's central coordinating office in Jakarta to provide bilingual manuals and visual aids for the trainees. Each Bechtel craft instructor was paired with an Indonesian instructor; the Bechtel trainers' jobs were to teach their counterparts to replace themselves as rapidly as possible.

During the course of these two construction projects, on-the-job training assignments were arranged for students from the cooperative program of the Engineering College of the University of Indonesia. At Bontag Bay six students were placed in field work situations that paralleled their formal engineering training. At Arun another group of students were given practical experience in such business skills as accounting and data processing.

*Algerian Liquefied Natural Gas Plant.* The world's largest LNG plant is located on the coast of the Mediterranean Sea at Arzew, Algeria. Bechtel's Refinery and Chemical Division made the original design study for the project, and later returned to the project to complete the EPC work initially undertaken by another contractor. It also had responsibility for managing the start-up and initial operation of the facility. A Bechtel team of approximately 150 was responsible for management during start-up operations and for the training of key client personnel to assume full operating responsibility gradually. During this period, Bechtel's supervisory personnel worked side by side with their counterparts at SONATRACH, the Algerian national oil and gas company, in a program designed to give the counterpart staff the greatest possible involvement and experience. In addition, an extensive classroom and on-the-job training program was carried out for O&M personnel.

### Advanced Technology Transfer

Where a project is of sufficiently long duration, and especially where the engineering and construction elements of the job are likely to be repeated in other applications over time, "advanced technology transfer" is possible. This means the transfer of technology and practical know-how experience and, importantly, the transfer of an understanding of the thought processes that go into technical and commercial decision making. To carry on this type of technology transfer, however, requires that the client or host nation have professional engineers available for the time-consuming classroom and on-the-job training that is associated with these programs. A major part of the know-how transfer is accomplished by having national participants actually perform elements of the project work.

Executing a complex project requires many technical disciplines and management skills. Bechtel uses a "project team" type of organization for its project work and encourages its clients to do the same for their new construction work. The project team is selected to incorporate the necessary skills required to handle the work and problems encountered in project implementation. The team is sufficiently small and compact to: ensure a shared understanding of the project goals by all team members; preserve good internal communication; and be sufficiently flexible to adapt to changing circumstances.

In its advanced technology transfer operations, Bechtel incorporates host nation engineers in its project teams both in the United States and in the host nation. Initial project conceptual engineering for the first project is most often done in Bechtel's major offices. This provides access to U.S. technology and current regulatory trends and permits Bechtel to assign at low cost a large number of well-qualified specialists to the project as needed to give it initial momentum.

Thereafter, if resources exist, the detailed design work can be carried out in the host nation. A team of Bechtel engineers accompanies the returning national engineers to lead the detailed design work to be performed by host nation engineers under the direction of the conceptual design team. Technical support from Bechtel's home office continues on an "as needed" basis.

An example of the distribution of Bechtel and host nation engineering manpower in a first generation nuclear power project involving advanced technology transfer is shown in Figure 4.4. In this project the client was committed to transfer of technology, and there were adequate trained national

**Figure 4.4. Power Project Total Engineering Man-hour Distribution.**

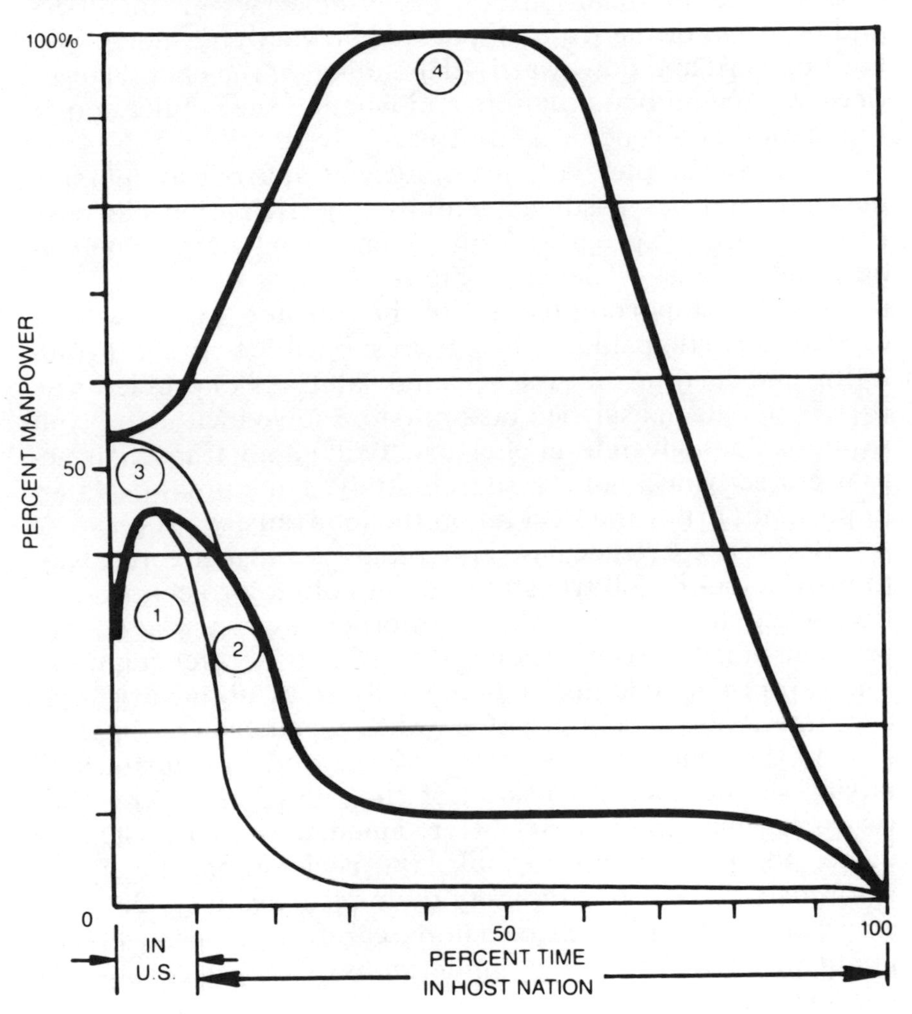


KEY

1 BECHTEL HOME OFFICE
2 BECHTEL IN HOST NATION
3 NATIONAL ENGINEERS IN BECHTEL HOME OFFICE
4 NATIONAL ENGINEERS IN HOST NATION

engineering resources available either from the client's staff or from other national sources. The number of team members and duration of the training period can vary considerably—both upward and downward—depending on the client's practices, the regulatory situation, and whether the duplication or replication of a generating unit is involved.

In this example, with availability of appropriate personnel and facilities assumed, national participation can rise rapidly as the job progresses and Bechtel participation phases out. This process is shown in Figure 4.5.

Another important feature of the transfer process is that each local participant is held fully responsible both for formulating and carrying out the selected solutions to problems as they occur. Each assigned task must be solved within a given time span and given level of effort. Technology transfer is actual project work, not a research study. Time pressure is an important part of the Bechtel on-the-job transfer process.

Bechtel experience has shown that the most effective way to provide on-the-job transfer and yet hold to project production schedules is to use the counterpart system: day-to-day relationships between national trainee and Bechtel engineer. The two project counterparts not only work physically adjacent to each other, but are also linked together organizationally. In the counterpart system, information transfer occurs during the detailed discussion of the "how" and "why" between assigned counterparts. Here again, positive transfer requires the participants to make the decision and defend it vigorously since it is ultimately their responsibility. Bechtel, however, monitors such decisions carefully to ensure that quality, cost, and schedule implications of such decisions are within prescribed limits.

During this process, roles and relationships develop and change as national engineers gain experience and confidence. The Bechtel role begins as supervisor, with a host nation trainee. In time, joint leadership is established, and gradually the principal leadership role transfers from Bechtel to the national. In subsequent projects of a similar nature, Bechtel eventually becomes a technical advisor to the national supervisor.

Figure 4.6 shows the complex interrelationships that exist in integrated project teams. Bechtel staff functions are shown

Figure 4.5. Percentage Participation in Design.

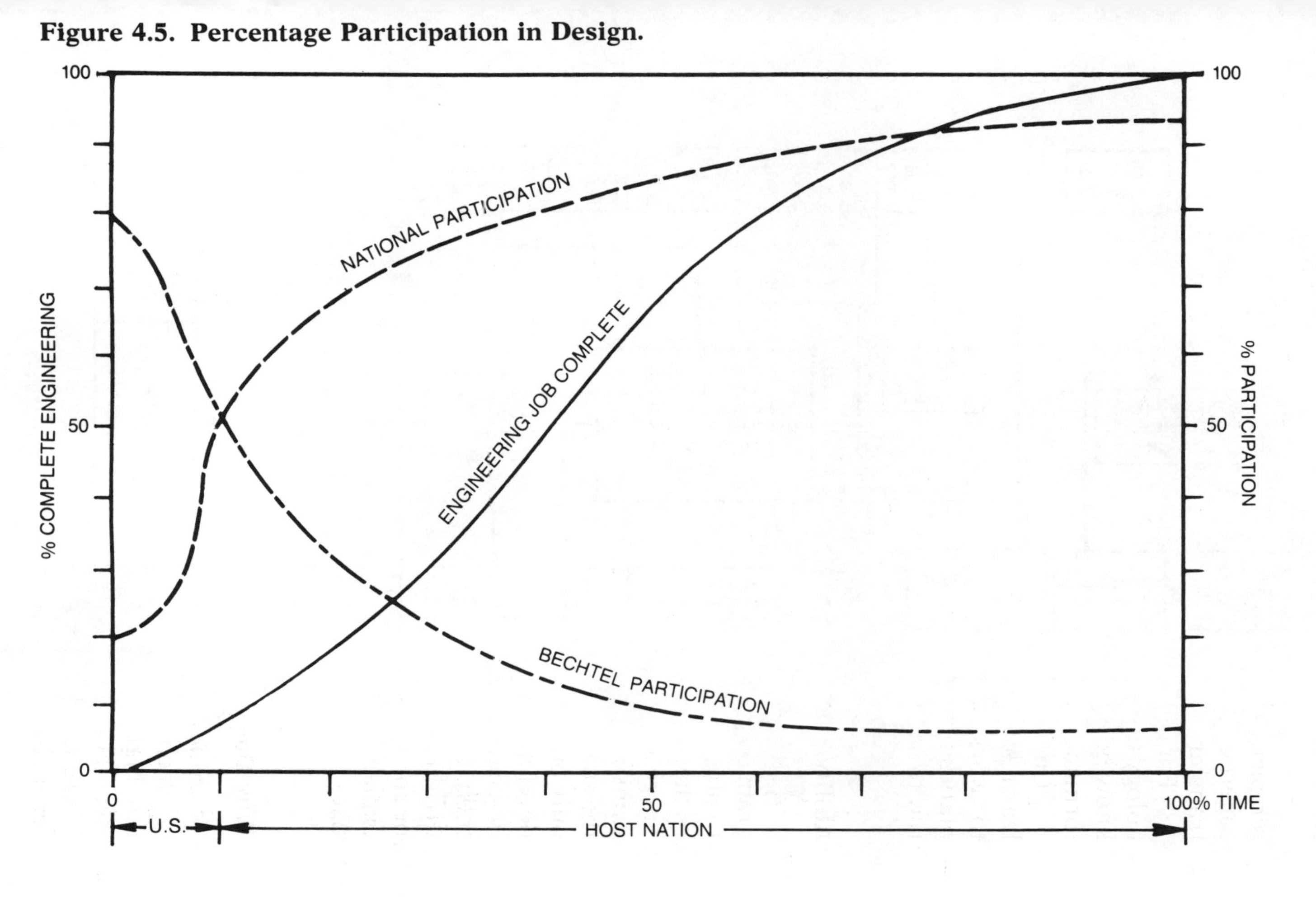

Figure 4.6. Typical Work Flow for an Individual Task.

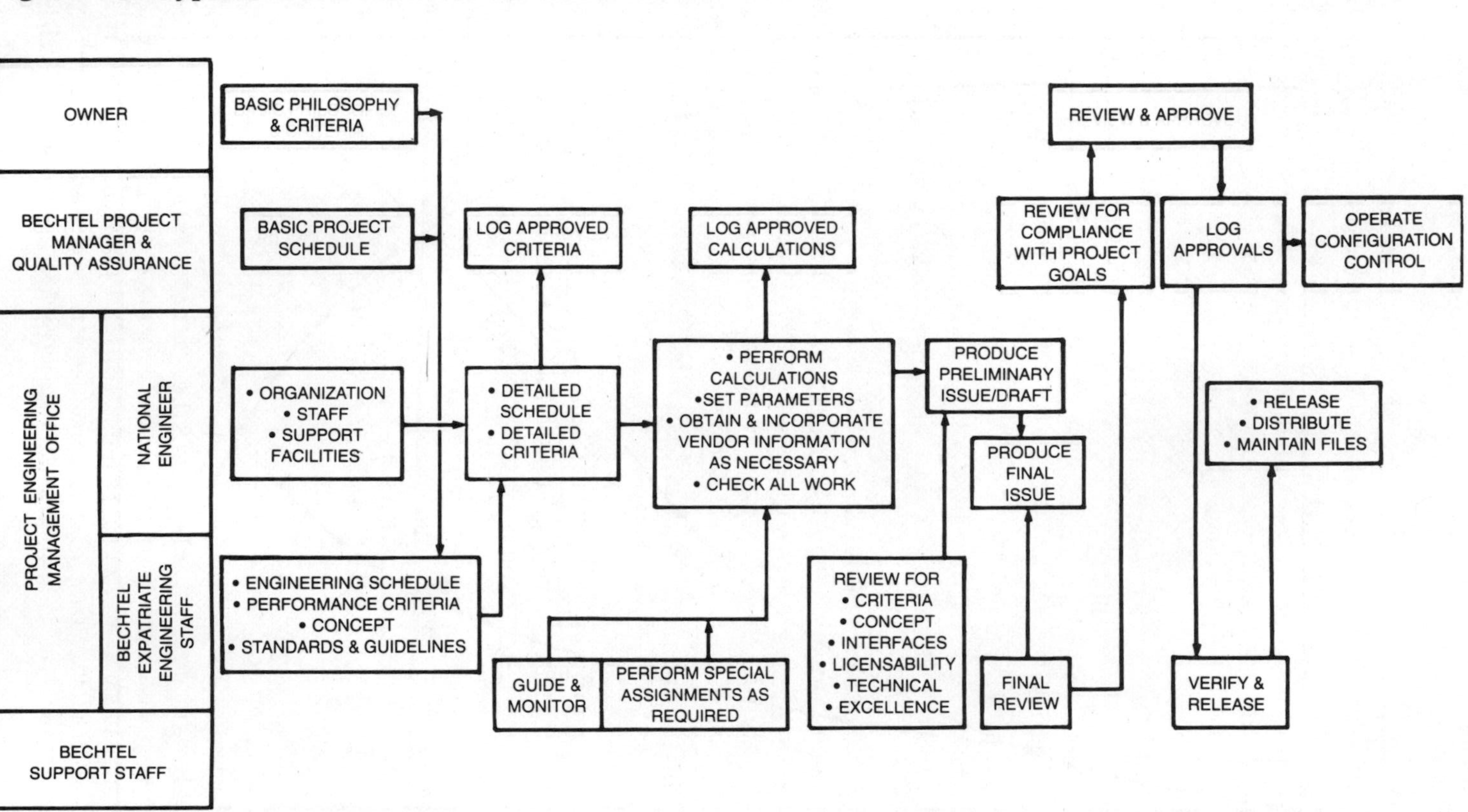

as shaded areas in the typical work flow of an individual engineering task, such as the preparation of a drawing, specification, or calculation during the detailed design phase in the host nation. The host nation client for the project sets or approves the basic philosophy and criteria for design, and reviews and approves the final design developed. National engineers do the actual engineering work under the guidance and with the assistance of Bechtel staff. Since Bechtel has the ultimate responsibility for the work, it maintains review and approval functions and schedule and quality control.

An example of Bechtel's engineering technology transfer is its work on the Metro system in Caracas, Venezuela. In this cooperative project, Bechtel's Hydro and Community Facilities Division is manager of the Parsons, Brinkerhoff, Tudor, Bechtel (PBTB) joint venture, which has designed and is providing construction management support for the Caracas Metro. The Metro will be a 56-kilometer, two-track rapid transit system with 55 stations and an estimated passenger volume of 760 million riders per year by the year 2000.

This project has an overall construction life of 30 years for its several phases. This long time interval has provided an opportunity for the gradual changeover of responsibilities from the joint venture partnership to Venezuelans. The project has moved from being engineered by foreigners to being designed, engineered, contracted for, and supervised by Venezuelans. Bechtel's own relationship to the project is changing from that of joint venture participant to partner with a Venezuelan engineering company that is increasingly assuming more responsibility and control.

Design work on the first phase of the Caracas Metro began in 1968. The basic system design work was done in the United States with approximately 40 engineers from Bechtel, 20 from Tudor Engineering, and 10 from Parsons, Brinkerhoff, Quade, and Douglas. All of these participants had been associated earlier on the San Francisco Bay Area Rapid Transit project and on other similar systems.

In 1969 the PBTB joint venture opened an office in Caracas to concentrate on the detailed civil design work in association with local contractors. That office grew to a peak staff of about 40 in 1970: 20 professionals from the PBTB joint venture and 20 civil design engineers from the Ministry of Public Works. The office worked closely with several local contractors in the civil and structural design of the Metro and in mechanical and electrical design work. In addition, several Ministry of Public Works engineers were resident in San Francisco working with Bechtel system designers during this period.

The evaluation of potential bidders for different parts of the project and, subsequently, of the bids received was a lengthy screening process. This was a period of major technology transfer from PBTB personnel to their Venezuelan counterparts through understanding of the bidding process and the criteria for selection of the winning proposal.

The PBTB foreign staff rose to about 50 by the end of 1979 in the construction management phase of the project. During this period the Caracas Metro Company was formally organized as an autonomous Venezuelan corporation apart from the Ministry of Public Works, and grew to over 600 employees, of which roughly 250 were engineers or technicians.

Bechtel's current role in partnership with a Venezuelan firm, which had been one of the initial subcontractors in the design phase, includes management advice in specialized subway construction disciplines, design services for the contractors building the line, project control assistance in scheduling, estimating and cost control, and assignment of personnel to the Caracas Metro Company to monitor the systemwide equipment design. The intercompany agreement between PBTB and its Venezuelan partner is based on PBTB's reducing its participation as the work progresses and the Venezuelan partner's increasing his. For the three-year period from the end of 1976 until the beginning of 1980, the Venezuelan partner's participation has grown from 8 professionals to 59.

## TECHNOLOGY TRANSFER BY THE BECHTEL POWER CORPORATION

The most comprehensive advanced technology transfer Bechtel has yet attempted is now taking place in the course of project work on several large, multiproject nuclear power programs. In addition to demanding engineering technology, nuclear power projects are extremely complicated, high technology construction and management jobs. Construction, for example, involves complex interrelationships between civil-structural elements and mechanical-electrical elements not only in engineering and procurement, but also during construction and erection.

Typical of the demanding aspects of nuclear plant construction are the confined working spaces caused by missile and radiation shields; limited equipment access portals; extreme weights that must be transported, lifted, and supported; complex wiring and instrumentation; and rigid welding standards. Equally demanding is the exacting quality of work required in nuclear equipment and materials.

Bechtel's current, large, multiproject nuclear power programs include three vital elements for successful advanced technology transfer: the use of the "components" approach to involve the maximum of existing host nation technical manufacturing, contractor, and managerial resources; multiproject involvement over an extended period of time to allow for the planned progression of responsibility for national professionals; and a formalized program for technology transfer both on the job and in the classroom, and a formal program for measurement of results.

## The Components Approach

Know-how transfer with respect to thermal power technology can be increased substantially if a client decides to develop its resources in the process of project implementation rather than utilize the lump-sum turnkey approach to contracting from foreign sources. Turnkey contracting in the power industry has come to mean that a manufacturer or consortium of companies does the design, supplies the equipment and materials, and carries out the construction and installation work of the project using equipment of its own manufacture. An engineering-construction company can be hired if the manufacturer or consortium does not have its own construction capability. In such turnkey contracts, most services are performed outside the country, except for the actual construction of the power plant, and even this often is done by imported labor.

To increase the opportunities for technology transfer, Bechtel uses what it calls the components approach in project implementation. This approach breaks down the project into work elements and contract packages that are tailored to utilize existing client and national capabilities without jeopardizing overall project control, which remains Bechtel's responsibility. An integrated project team, consisting of Bechtel, client, national engineers, and others, does the detailed engineering work for all systems, develops specifications for equipment, and hires and manages local contractors for erection and construction of the plant.

Using this approach to project implementation, the local utility company is actively engaged in design, procurement, and erection, and is in total control of all key decisions. These are basically new activities for a utility using the components approach for the first time. Typically, several hundred equipment specifications, with anywhere from 6 to 12 construction and installation contracts, are involved. A client's decision to utilize the components approach requires a large commitment of its resources and its management attention. On-the-job and classroom training programs are used to assist the transfer and implementation.

A graphic representation of participants, disciplines, and the scope involved in a typical components approach project is shown in Figure 4.7. The integrated project team, in effect, provides the client-owner with an extension of the staff for as long as required. This arrangement permits the integrated Bechtel-client team to operate on the client's behalf in achieving the desired goals; the arrangement can yield broad and thorough exchange of technology and know-how among team members. The structure is highly flexible; it can be adapted to permit maximum use of existing local resources. The ability to separate procurement contracts from construction contracts facilitates the use of local sources of supply and allows the maximum use of competitive sources for equipment and materials, yet maintains standards.

The project production organization that carries out the components approach to nuclear power plant construction usually consists of four major components:

**Figure 4.7. Components Approach Project Structure.**

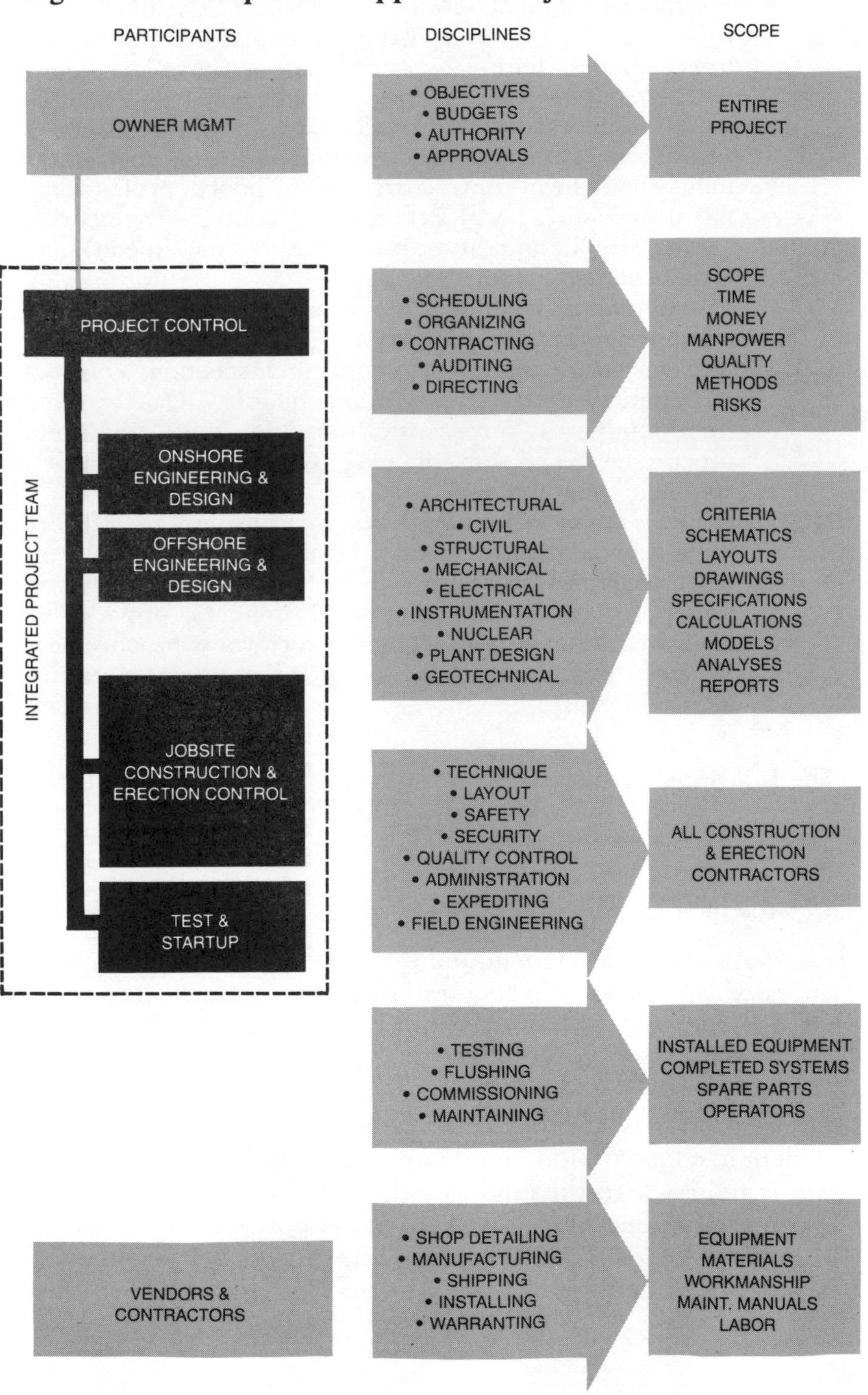

PARTICIPANTS
DISCIPLINES
SCOPE
OWNER MGMT
• OBJECTIVES
• BUDGETS
• AUTHORITY
• APPROVALS
ENTIRE PROJECT
INTEGRATED PROJECT TEAM
PROJECT CONTROL
• SCHEDULING
• ORGANIZING
• CONTRACTING
• AUDITING
• DIRECTING
SCOPE
TIME
MONEY
MANPOWER
QUALITY
METHODS
RISKS
ONSHORE ENGINEERING & DESIGN
OFFSHORE ENGINEERING & DESIGN
• ARCHITECTURAL
• CIVIL
• STRUCTURAL
• MECHANICAL
• ELECTRICAL
• INSTRUMENTATION
• NUCLEAR
• PLANT DESIGN
• GEOTECHNICAL
CRITERIA
SCHEMATICS
LAYOUTS
DRAWINGS
SPECIFICATIONS
CALCULATIONS
MODELS
ANALYSES
REPORTS
JOBSITE CONSTRUCTION & ERECTION CONTROL
• TECHNIQUE
• LAYOUT
• SAFETY
• SECURITY
• QUALITY CONTROL
• ADMINISTRATION
• EXPEDITING
• FIELD ENGINEERING
ALL CONSTRUCTION & ERECTION CONTRACTORS
TEST & STARTUP
• TESTING
• FLUSHING
• COMMISSIONING
• MAINTAINING
INSTALLED EQUIPMENT
COMPLETED SYSTEMS
SPARE PARTS
OPERATORS
VENDORS & CONTRACTORS
• SHOP DETAILING
• MANUFACTURING
• SHIPPING
• INSTALLING
• WARRANTING
EQUIPMENT
MATERIALS
WORKMANSHIP
MAINT. MANUALS
LABOR

- *Project control and contracting:* A Project Control Office is responsible for budgeting, scheduling, planning, forecasting, soliciting bids, negotiating contracts, policy making, and overall reporting.
- *Onshore engineering and vendor monitoring:* The Onshore Engineering Office in the host nation provides ready access to the client, local contractors, and the pool of professional talent available. Local Bechtel staff controls engineering quality, supplier quality surveillance, and the expediting of local vendors.
- *Job site contract administration and start-up*: The typical job site manager is responsible for detailed engineering, construction planning and control, inspection, contract formulation and administration, quantity tracking and forecasting, cost forecasting, schedule monitoring, field procurement, and surveillance of quality, safety, security, and labor relations.
- *Offshore engineering and vendor monitoring:* The Offshore Engineering Office is located inside Bechtel's permanent power engineering offices where access to top staff and project records is easy. Functional support is provided by divisional chiefs, and Bechtel's worldwide procurement support office controls the supply of imported materials and equipment.

The relationship of these four offices within the overall project structure is shown in Figure 4.8.

## Multiproject Involvement

More than a single isolated project is required to transfer the necessary advanced technology and project management skills needed for complex installations. Effective transfer requires multiyear, multiproject, multidiscipline involvement. Performing the act of design engineering is an incomplete experience; only when the result of such engineering becomes evident in construction, plant start-up, or in routine operation and is fed back to the engineer does the learning experience become complete. An integrated series of projects spaced over perhaps a decade is needed to provide this type of learning experience for nuclear power plants. Such an extended program

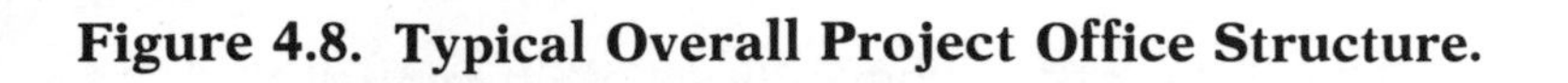
**Figure 4.8. Typical Overall Project Office Structure.**

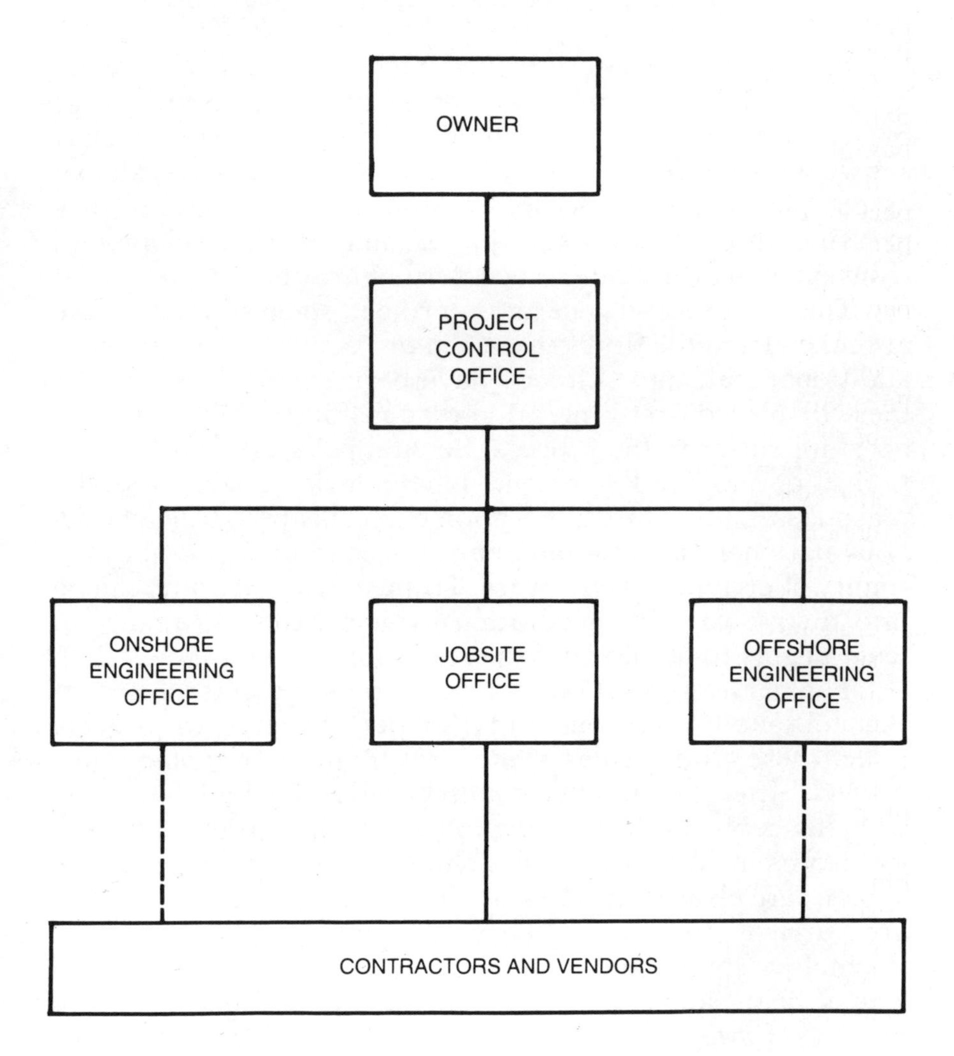

allows for the planned progression of responsibility to national engineers and professionals in project engineering and management.

Figure 4.9 shows the involved interrelationships that exist in multiproject programs. Four overlapping "generations" of projects are shown as initiated during the time period (each is shown with a different line code). One of the projects is shown as completed during the period shown in the diagram, progressing through six overlapping phases of development, from planning through to successful start-up.

To staff a second generation project, some trained people are taken from the first generation project. As a consequence, additional staff must already have been trained to take over these displaced first generation responsibilities, so that work does not suffer in the process. Bechtel pays particular attention to developing key people, particularly supervisors. The reason is that the third generation projects are to be led largely by nationals. Careful planning is needed to ensure the continuity of committed engineers. Trainees, for example, can be into their own third generation project before the start-up team begins to obtain any experience on the initial one.

For first generation projects, most clients ask that Bechtel take full responsibility for performance. In projects under these ground rules, some local engineers are placed in a project office in the host country under Bechtel guidance, whereas certain other key nationals are integrated into project teams in an established Bechtel power engineering office somewhere else. At the job site, first generation projects often are carried out with national participants supporting key Bechtel personnel and with essentially all other job site personnel being local nationals. Each successive generation of projects, however, involves fewer and fewer Bechtel professionals.

In second generation projects, for example, key project roles are filled with national engineers wherever possible, with Bechtel acting as personnel supplier of last resort. Usually, all but initial systems engineering work can be carried out in local project offices with Bechtel support.

In third generation projects, nationals normally are assigned all key roles, assisted by Bechtel. All project work is

Figure 4.9. An Example of a Multiphase Nuclear Transfer Program.

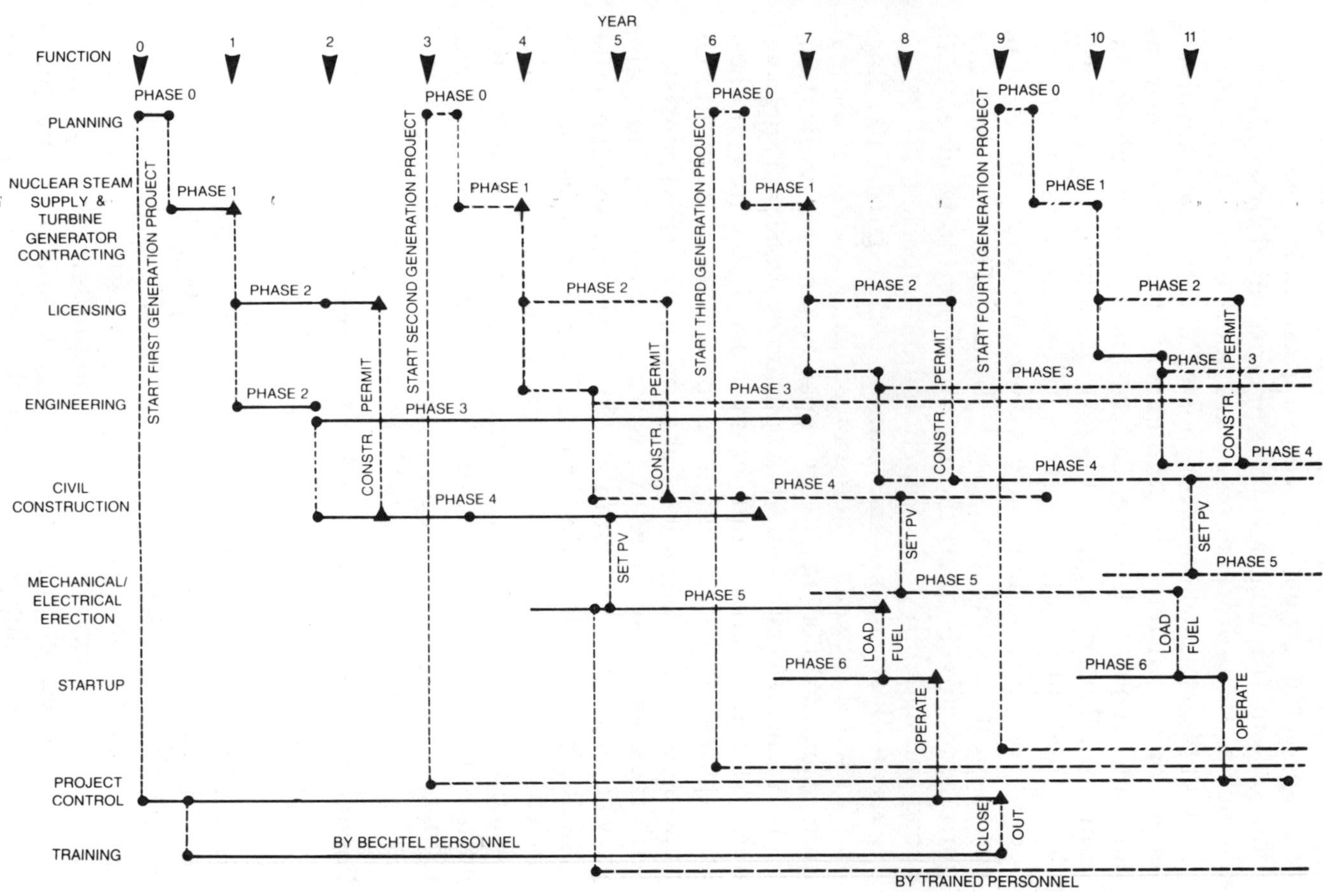

done in local project and field offices. For fourth generation projects, only occasional advice, some project control support, special investigations, and supplementary specialist and management training normally are provided by Bechtel inasmuch as local professionals have taken over.

For multiproject nuclear power assignments involving technology transfer, the number of trained engineering personnel required is large. Except for the initial phase, the number of Bechtel engineers involved is small; the great bulk of the total supply of workers of all generations of projects comes from host nation sources—whether from the client or from local architect-engineer firms. The training requirements are complex; those trained need to be given both broad and in-depth project participation experience.

As an illustration of the training task, Figure 4.10 shows that with a three-year spacing of projects with twin nuclear units each of 1,000-megawatt capacity (and no replication), there is a need to build up a host nation professional engineering and construction staff of at least 1,300 to run these overlapping projects. Operating personnel are additional. Spacing between generations of projects can vary, consistent with the client's power needs; worker availability must also vary accordingly. For example, with a five-year spacing of the generations of projects, fewer than 1,000 people would be involved, but with large swings in personnel levels; this causes difficulties in personnel continuity. Under any spacing, the number of people needed is still large, however, if meaningful technology transfer is to take place.

## Formalized Technology Transfer Programs

With the heightened appreciation of technology transfer as a vital aspect of international assignments, the Bechtel Power Corporation has formalized a Technology Transfer Organization to facilitate long-range, planned transfer of engineering and construction technology. The essential functions of the Technology Transfer Organization include: planning and carrying out formal training programs designed to teach technical engineering and project management; planning and facilitating long-term, on-the-job training programs for participants, designed to provide a planned series of responsibility changes from Bechtel to recipients of the transfer; interfacing with other groups in the client's organization and in Bechtel in the coordination of technology transfer activities, in the evaluation of its program, in the development and upgrading of course material, and in the development of new programs. A simplified organizational chart of this program is shown in Figure 4.11.

Formal training in a technology transfer program extends over the life of the project. In the most comprehensive program, the first year places a heavy initial emphasis on classroom training, with up to one-quarter of a participant's available time spent in course work prior to entering project work. Thirty to forty technical courses may be involved, supplemented by instruction in oral and technical English. All courses are taught by practicing Bechtel professionals from active projects. The remainder of the participant's time is spent in on-the-job training in actual project work.

In the second year, formal training drops down to about one-tenth of available time, with the remainder in on-the-job training in project work. Thereafter, formal classroom technical training stabilizes at an average of 5 to 10 percent of available time. As of mid-1980, the Los Angeles Power Division was using about 125 classroom courses, with another 200 to 300 courses under development. Some of these courses are also available in Spanish, Korean, and Chinese.

The participant's on-the-job training is a key element in the transfer of technology and know-how by Bechtel. The participants become active working members of a project team, have work assigned to them, make decisions, take responsibility, and work under the time constraints of production schedules. Supplemented by off-the-job study courses as required to achieve planned goals for each participant, the high pressure "learning and doing" project work environment is an essential element of the transfer process. Working on an integrated project team also offers an opportunity for a surprising amount of learning by "listening"—Participants are exposed to the reasoning process of experienced engineers in making the decisions they do and learn more practical ways to carry out specific work assignments by the very simple mechanism of proximity.

**Figure 4.10. Estimated Manpower Requirements (based on twin 1,000-MWE nuclear units).**

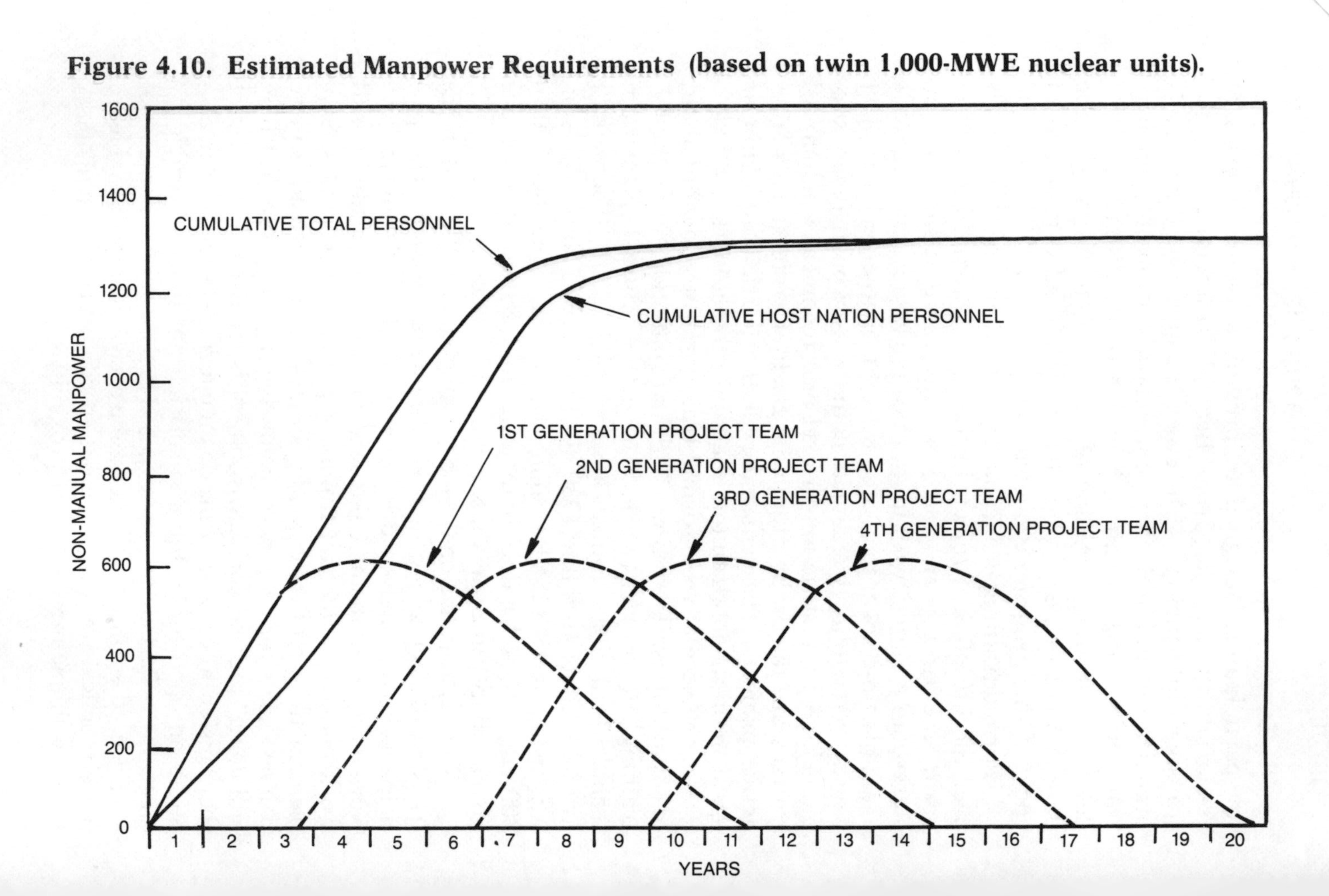

Figure 4.11. Technology Transfer Operations Organization.

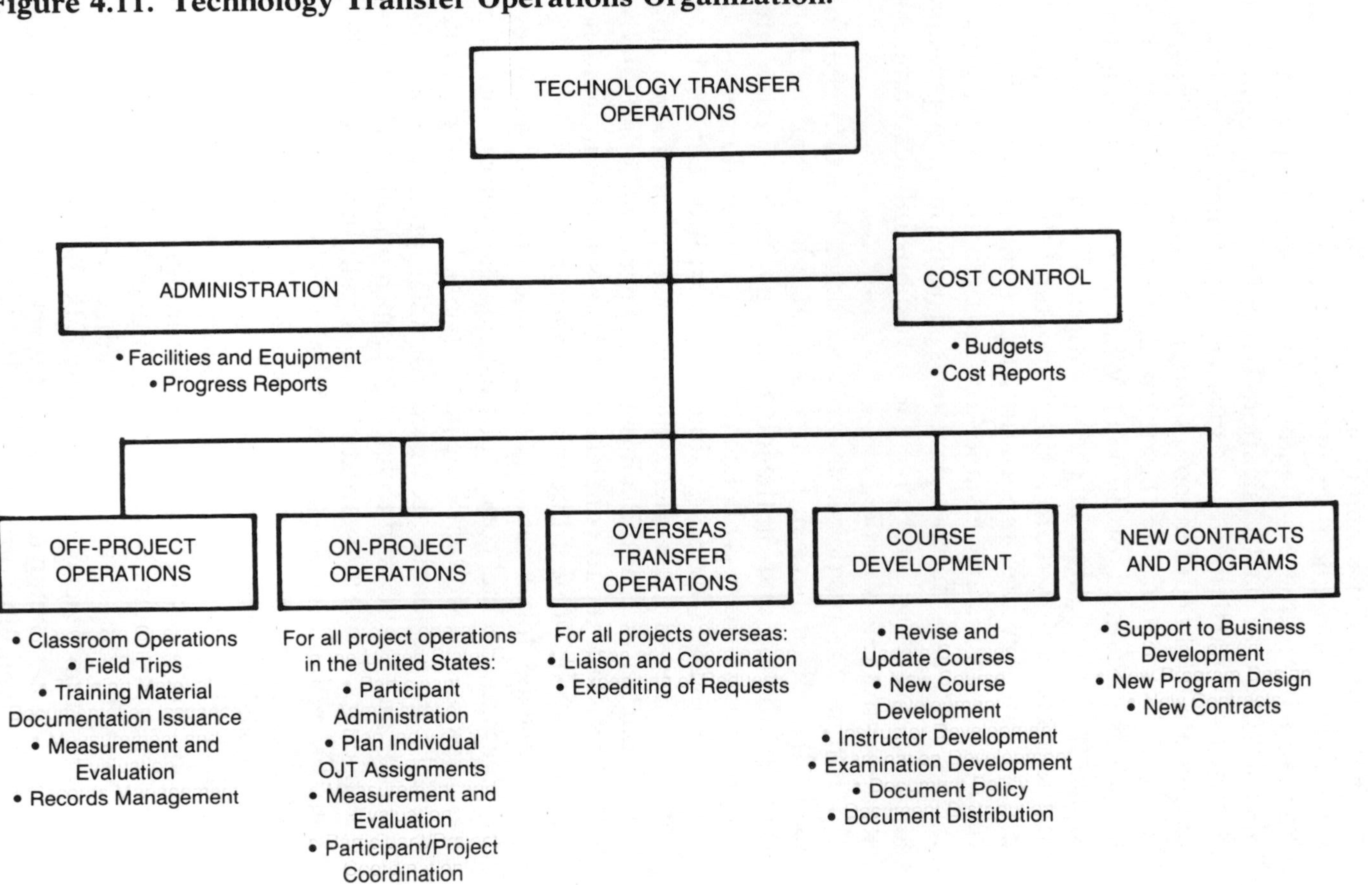

Participants' training programs are tailored to the objectives they seek in their long-term professional development. The tailoring of the off-the-job study and on-the-job training also depends on past educational accomplishments and experience. All participants are given a comprehensive entrance examination (almost five hours long) aimed at the job profiles for which they are to be trained; that is, categories of knowledge required for the job ("What is it he or she ought to know?") are weighted in the questionnaire design. The absolute examination score is not important; the scores in each profile category are important. After evaluation of the test results, programs are designed for participants to strengthen their weaknesses, further develop strengths in their primary areas of expertise, and give them responsibilities commensurate with their abilities.

Thereafter, participants are periodically examined for evaluation of changes in their knowledge as a result of both their off-the-job and on-the-job training. The on-the-job productivity and progress of each participant are evaluated about every six months by an Evaluation Committee consisting of representatives of the client, the recipient group responsible for the participant, and Bechtel project personnel, and also by a Bechtel staff group consisting of chiefs and managers not directly on the project but responsible for it.

A driving force behind positive and efficient technology transfer is the Bechtel policy of measuring the performance of its own personnel as well as of the participants. In transfer projects, Bechtel's professionals are rated for their project production schedule and quality and technology transfer quality, depth, and effectiveness. These two yardsticks are usually in serious conflict; that is why both are measured. A balance is always required: Both efforts are deemed to be important. The careers of Bechtel professionals depend upon their performing well in both.

Bechtel counterparts in on-the-job training programs are rated by the work completed by the integrated project team ("Is it of high quality? on time? on budget?") and, on a one-to-one basis, by the achievement of their counterpart trainee in the evaluation described previously ("Is the participant gaining knowledge? being a productive worker? accepting responsibility? developing on a career path?").

Bechtel's instructors (practicing project professionals) and courses in the formal classroom training programs also are rated both by their students and by senior professionals.

**Examples of Experience**

The following examples of Bechtel's experience in thermal power technology transfer show the evolution of the transfer process.

Perhaps Bechtel's earliest contract for conventional thermal power technology transfer in a rapidly developing economy came in Korea in the 1950s. On-the-job training was an integral part of the program, and many of the engineers involved in these programs have become top managers in Korea.

Since the 1960s Bechtel has had assignments for a variety of clients in Mexico emphasizing technology transfer. Characteristic of these assignments is the phase out of Bechtel participation over time while leaving an independent operating organization with the capacity to design and construct thermal power projects.

The components approach was first adopted as a technology transfer tool by Bechtel in Spain to carry out a series of related nuclear power plant projects. Much of the engineering and construction capability needed for maximum benefit to be derived from the components approach was already in place from hydro and fossil fuel plants. After roughly eight years, Bechtel is now in the phase out of its involvement in these early projects, but is continuing in a supervisory role in second and third generation work.

In Taiwan, Bechtel has moved to the formation of a jointly owned engineering and construction company based in Taiwan initially to accelerate and broaden the transfer of technology and professional know-how in all phases of Taiwan's ambitious nuclear program. This company is now in the process of establishing a Technology Transfer Operations group of its own patterned on the model developed in the United States.

*Korea*

In 1954 Bechtel began design and construction of three thermal power plants in South Korea. These plants doubled the country's available power supply in less than two years. One of the most vital aspects of the job was the hiring and training of Koreans at all levels of responsibility. There were fewer than 100 U.S. personnel at each of the centers of the construction program, but more than five times that number of Koreans. On-the-job training was an integral and indispensable part of the program.

Early in the project, recruitment began for 42 Koreans to be trained in the United States in the technical phases of modern steam power plant operations, so that they could take over operation of the new plants on a permanent basis. Their eight-month training program included: a special curriculum developed by Bechtel and carried out by the University of California at Berkley; a practical experience training program in construction and operation of electric plants; and familiarization experience at companies manufacturing the principal equipment for the power plants that were under construction. Once back in Korea, these engineers acquainted themselves with and studied operational procedures for the new plants. They also served as instructors, holding classes for local trainees selected from the most promising members of the construction force. These trainees later became their assistants. The job was completed ahead of schedule and under budget.

Currently, Bechtel is providing design EPC management services as well as training of the professional staff of KECO and its associated engineering company, Korean Nuclear Engineering, in support of KECO's major nuclear power program. An integrated project team of Bechtel and Korean engineers is currently carrying out the design of four major nuclear generating units. As part of that team effort, an office has been established in Seoul to oversee local vendors and a job site office set up to oversee construction contractors for these units. Some 100 Koreans are actively participating in the U.S. portion of the production effort, through work in joint project teams in formal technology transfer programs. Major programs of know-how transfer in construction techniques, field engineering, quality control, and construction management are also being carried out at the job site for KECO and local Korean contractors.

Over 400 Koreans are currently involved in formalized training and project work with Bechtel, and this number is being expanded substantially.

In addition to these nuclear programs, transfer work is underway on five oil-fired power plants and two coal-fired plants. These latest coal-fired units are being designed in Korea using integrated project teams, following a short, initial design and transfer phase in the United States. In its coal-fired program, Bechtel is providing its services to KECO through a joint venture company formed by Bechtel and a Korean engineering firm. Practical experience in project work is supplemented in Korea by classroom training programs.

To date, Bechtel estimates that 800 Korean engineers and technicians have been trained on project work, and, in addition, Bechtel has carried out a number of intensive six-week programs in project management techniques and systems for KECO project managers and senior management.

*Mexico*

In the early 1960s Bechtel undertook a consulting assignment for the Comision Federal de Electricidad (CFE), the national utility responsible for providing Mexico's electrical power, to determine the best approaches to the conversion of the electric power system to a uniform frequency standard throughout the country.

Technology transfer was involved in all phases of the work that resulted from this initial assignment. Even in the original consulting study, a joint team approach (consisting of seven or eight Bechtel engineers working with a like number of CFE engineers) was used in the gathering of data, for the analysis of alternatives, and in the formulation of recommendations. That same joint team worked late in the second phase of work that resulted from the initial study, namely, detailed design and planning of the actual conversion effort. The second phase work was, in effect, the development of a master plan for the several-year implementation program, adapting Bechtel's experience from similar assignments in the United States and Venezuela to Mexican conditions.

When the third phase work—the actual frequency conversion program—was about to take place, a new organization, independent of CFE, was created to bring about the frequency changeover. The same joint Bechtel-CFE team that had worked on the earlier phases became the nucleus of a new organization that grew to 1,500 personnel, about 500 of whom were engineers. The new employees were young, inexperienced people recruited and selected from schools and universities. For these new employees, the joint team developed training courses and practical experience programs in the industries that would be involved in the frequency conversion. The design and planning manuals developed in the design phase of the assignment became the basis for the training program for new employees.

Further experience was gained from pilot conversions of small towns and pilot industries in Mexico City before the full scale program began. The initial new employees in time became trainers of subsequent new employees. Bechtel's role in the organization gradually became less over the three-year life of the organization, as the technical and managerial know-how for frequency conversion was successfully transferred.

Although most of the employees of the newly formed organization did not become CFE employees after this "one-time" conversion work was completed, they entered Mexican industry as experienced engineers, having benefited from the technical and know-how transfer that took place in the course of the Bechtel project.

In 1968 one of Mexico's largest engineering and heavy construction companies, Ingenieros Civiles Asociados, and Bechtel formed a joint company in Mexico to carry out power and other industrial projects. Initially, 25 Bechtel engineers went to Mexico to work with 80 engineers from the Mexican engineer-constructor. They jointly formed an organiztion that grew to about 150 engineers in the course of design work for a total of 2,000 megawatts of fossil-fueled electrical generating facilities for CFE. Toward the conclusion of this joint effort, U.S. resident professional support in the joint company phased out, and the number of Mexican engineers carrying out the continuing work increased to 120. Subsequently, the work was transferred entirely to the Mexican engineering firm.

In 1976 CFE decided to assume responsibility for the engineering, management, and construction of its ambitious fossil-fueled expansion program, which included 11 new hydroelectric units, 16 new thermoelectric units, and the country's first nuclear power station. A new Department of Engineering and Design was set up within CFE to be used for training its own engineers, designers, and drafters for this expansion program. CFE requested Bechtel to transfer technology and know-how to the staff of this new department in the course of the actual design and engineering work for 5 of the new stations (11 generating units having a capacity of 2,100 megawatts).

In addition to serving as technical advisor to CFE on the design and engineering of the power units, the contract for this work explicitly defined some aspects of the transfer of technology ("technical knowledge") to be included in the work. For example, Bechtel provided the Comision with a virtual library of manuals, bulletins, and guides for engineering construction management activities, as well as access to Bechtel's project control programs including computer programs. The training methods and personnel to be involved in them are shown in Table 4.1.

**Table 4.1. Training Methods and Personnel in Transfer of Technology to CFE**

| *Phase* | *Method* | *Source* |
|---|---|---|
| Basic orientation | Lectures, films, questions and answers | Bechtel training specialists and consultants |
| Basic design and off-the-job training | Lectures, textbooks, quizzes, programmed instruction, films, diapositives | CFE supervisors and Bechtel consultants |
| Supervisory training | Films, problem solving, lectures, texts | CFE supervisors and Bechtel consultants |
| On-the-job training | Specific practical work assignments, coaching, performance appraisal, counseling | CFE supervisors supported by Bechtel consultants |

Once again a joint team approach was used by Bechtel and CFE to carry out this work. After about three months of planning by the joint team, recruitment of new staff began. Within three months the new employees totaled 200, and by the end of the two years increased to the present 400 in the Mexican design department. Bechtel resident staff peaked at about 50 during the course of this five-year assignment.

The technology transfer program was intensive. It involved engineers, drafters, supervisors, schedulers, and cost control professionals. The transfer was effected by on-the-job participation in all activities concerned with the design of the 11 new power-generating units. This "learning by doing" was supplemented by an extensive program of classroom training. The initial, practical steam power plant design course was followed by transfer programs in all engineering disciplines and in specific engineering specialties. A total of almost 50,000 person-hours of CFE employee time was invested in classroom transfer programs, in addition to over 1 million person-hours in on-the-job project participation.

The technology transfer program for this new CFE thermal power design and engineering department took about four years. By the end of that time, the transferors had worked themselves "out of a job." Although in the beginning, classroom programs were run primarily by Bechtel engineers, within three years CFE engineers were doing it themselves. The Mexican professionals involved in the technology transfer program have now assumed full responsibility for the new plant additions to be made to the CFE system. Three new power stations designed and engineered in the course of this technology transfer program are already built and successfully operating.

The director of the CFE design and engineering department is extremely satisfied with what has been accomplished in this technology transfer program. His prime design group staff, he feels, has the same quality as the Bechtel professionals who were involved in the transfer process. All the roughly 400 new employees trained in the course of this program remain productively employed within CFE; yet there are fears that in time some of them will be hired away by other Mexican firms because of their high professional quality.

As further evidence of the successful transfer of technology that has occurred as a result of this program, most architect-engineer firms practicing in Mexico are said to be using copies of the project operating procedures developed for CFE in the course of their own professional work.

With fossil fuel thermal power plant technology now transferred, Bechtel is assisting CFE in the creation of a nuclear design and engineering group and in overall management control systems for the improved planning and control of all CFE projects.

In the project management area, as of 1981, ten thermoelectric power projects in progress had been chosen for on-the-job training of a new group assembled from existing CFE employees. Ten project teams had been set up—one for each of the plants. A pilot project team was headed by a Bechtel manager who has the responsibility for the team's effort in budgeting, cost control, and scheduling. The pilot group was serving as an instruction vehicle for the other teams in the group's program. The other teams were headed by CFE engineers supported by Bechtel consultants. Other Bechtel engineers support staff functions set up by the new group in cost control, scheduling, and systems development. In this total effort, 10 Bechtel engineers were working with a group that had grown to 35 professionals.

*Spain*

Beginning in late 1971, working closely with a number of Spanish utilities and local engineering and construction firms, Bechtel participated in a series of nuclear power plant design projects utilizing the components approach abroad for the first time. The first project consisted of two nuclear units of the Lemoniz power station near Bilbao for Iberduero, Spain's largest electrical utility. Initial engineering design work was done by Bechtel in the United States; then, Bilbao and Madrid project offices were established with a mix of Bechtel and local Spanish engineers under joint Bechtel-Iberduero leadership. Bechtel know-how was introduced by Bechtel engineers assigned to the various disciplines groups. In the course of this project, up to 200 Bechtel professionals have worked in Bilbao on assignments ranging from months to six years.

On the second generation project for Iberduero and two other utilities—one nuclear unit each at Vandellos and Sayago—approximately 30 Spanish engineers, who had previous fossil design experience, worked with Bechtel engineers in the United States to develop the initial plant designs. After approximately nine months, the team for each project was able to complete basic site and plant arrangements, systems definition, safety analysis reports, and purchase orders for major plant equipment. Each team then returned to Spain, accompanied by about 30 Bechtel counterpart engineers, to complete the detailed designs of the plants and manage the project. Three Spanish engineering firms were assigned responsibility for portions of the detailed design.

The supervision and leadership of each project discipline (mechanical, nuclear, civil, electrical) were performed by a Bechtel engineer and Spanish counterpart. This one-to-one relationship extended to all the project disciplines required.

A similar form of working relationship was established with five other utilities for two nuclear units in the ASCO project southwest of Barcelona. Bechtel's participation has included management of the project engineering office consisting of Bechtel engineers and those from two Spanish engineering firms. Some 270 engineers and drafters of those firms have been involved in the design work. Bechtel supports the client utilities in project management, planning, and scheduling, cost estimating and control, quality assurance, and procurement and construction. A Spanish work force of 3,100 is involved in the actual construction of ASCO.

The technology transfer program has so evolved that today all work is performed in Spain by Spanish engineers with relatively few Bechtel professionals in that country. Nearly 2,000 Spanish technical and professional personnel have been involved in this technology transfer program, wherein host nation engineers and managers have taken over positions of responsibility, gradually displacing Bechtel involvement in the projects.

*Taiwan*

Bechtel is assisting the Taiwan Power Company in carrying out its major nuclear power program. Currently, the program involves the engineering design and construction of four nuclear units, procurement, technical management of construction, engineering supervision in Taiwan, and start-up of these units at two job sites—Maanshan and Kuosheng.

Taiwan Power is handling its own construction with Chinese contractors, and thus the professional job site Bechtel staff has been small. At peak, at the Maanshan job site, for example, Bechtel's staff will not exceed 50 employees. This figure contrasts with an expected peak of about 3,500 employed by Taiwan Power and its subcontractors. The small Taiwan job site Bechtel staff is primarily advisory; its role is technical management, direction, and training. The program has included assistance in the training of contractor personnel. The engineering, design, and procurement activities for the two Maanshan project reactors, however, have required a peak staff of about 250 persons in the United States.

As part of its program with Taiwan Power, Bechtel has trained many Chinese professionals in engineering, construction, quality control, procurement, start-up, planning, and management. The training took place both in the United States with the Bechtel project team and later at the job site in Taiwan. In the spring of 1981, 30 Bechtel engineers in Taiwan were supervising the detailed design work of 320 Chinese engineers and drafters for the Maanshan and Kuosheng projects.

The use of the components approach in Taiwan has been supportive of the government's program to achieve maximum use of local personnel, materials, and manufacturers. Bechtel is providing liaison services to local firms supplying construction materials and equipment. Several joint ventures with foreign companies (not Bechtel) have been established in Taiwan for the manufacture and fabrication of plant equipment for nuclear units, including turbine generators, components of nuclear steam supply systems, piping, pipe hangers, valves, cable, condensers, tanks, pumps, and fabricated steel.

Recently, Bechtel and Taiwan's Sinotech Engineering, Inc., formed a new, jointly owned engineering and construction company based in Taiwan—Pacific Engineers and Constructors, Ltd. The initial purpose of the new company is to accelerate and broaden the transfer of technology and professional know-how in all phases of the next phase of Taiwan's nuclear program.

Each of the nuclear units scheduled for the second phase of Taiwan Power's program will take about six to eight years to complete, with a capital investment of approximately $1.3 billion each. Of that figure, about 50 percent will be spent on equipment and materials. During the construction phase, employment will be between 3,000 and 4,000 at each site.

Bechtel technology transfer efforts for Phase I of the Taiwan Power nuclear program have to date resulted in the training of an estimated 1,000 Chinese professionals.

In summary, it is estimated that over the last 25 years, the Bechtel Power Corporation has trained some 10,000 technical, professional, and nonmanual personnel in the course of its international power projects. The cumulative number of those so trained and the number of Bechtel personnel involved at any one period of time are shown in Figure 4.12.

## A HOST NATION'S VIEW OF BECHTEL'S TECHNOLOGY TRANSFER ACTIVITIES

To obtain an appreciation of the value of Bechtel's services as seen by clients, a limited number of interviews were held in Mexico City with several of Bechtel's clients, a joint venture partner, representatives of the government, and professionals in academia. Their views are summarized here after a brief review of Mexican government regulations regarding technology transfer.

### Background

Mexico is a country that has been fulfilling its internal demand for technology largely through imports of foreign technology, importantly from the United States. Research undertaken for the Mexican government by national and foreign economists in the late 1960s looked into the form in which technology was being imported and the payment mechanisms for it. This study led to a 1972 Law on the Registration of Transfer of Technology and the Use of Exploitation of Patents and Brands. Although essentially flexible in its administration so that the flow of technology will not be interrupted, the law does prohibit a number of restrictive provisions in such contracts.

**Figure 4.12. Bechtel Power Technology Transfer Experience.**

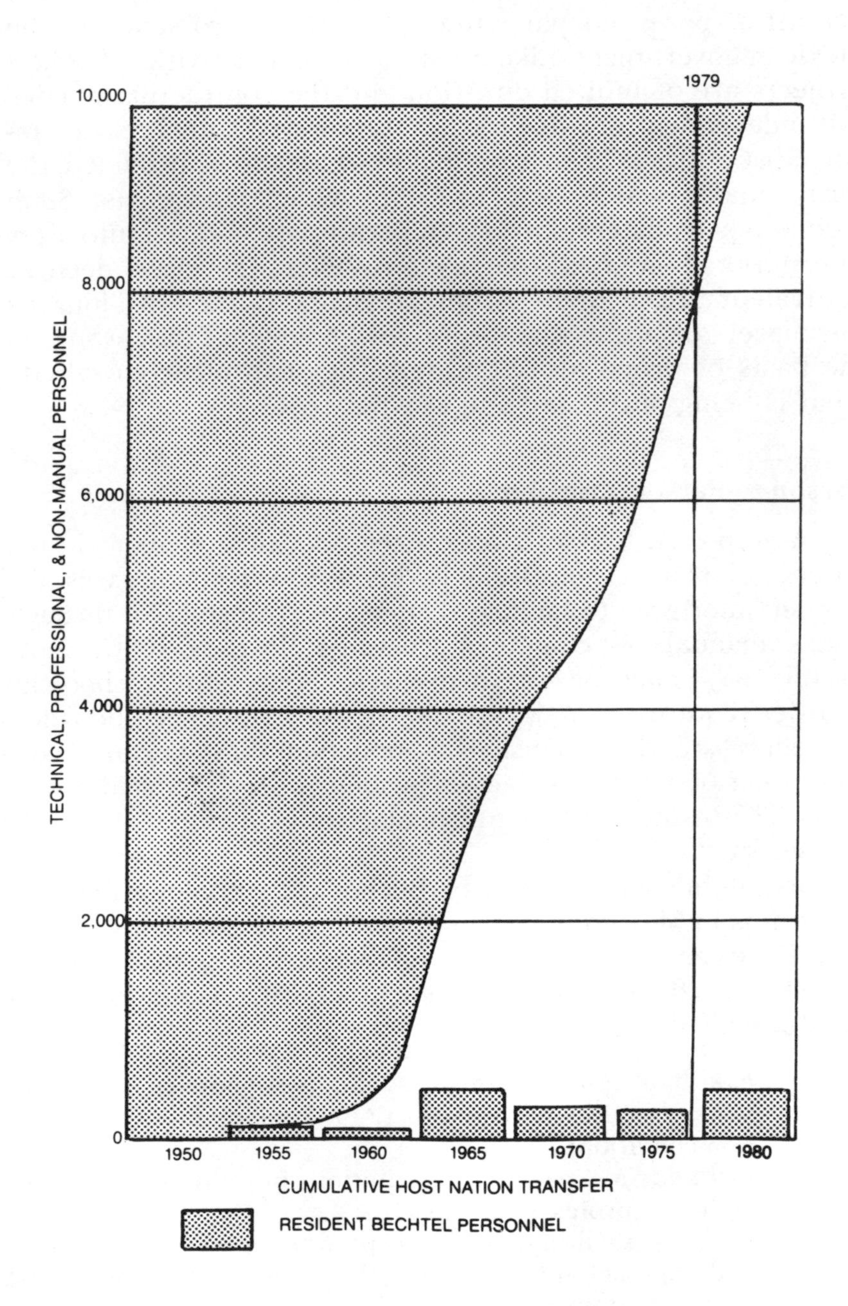

Transfer of professional services technology, as practiced by Bechtel in its engineering, construction, and management activities, poses no particular regulatory problem for the Mexican government. Like most consulting activities, Bechtel projects are of limited duration and the contracting parties are independent entities (no intercompany transfers are involved). Contracts clearly delineate what is to be provided, the terms under which it is to be provided, and costs. Such technology transfer contracts generally are entered into after international competition, except where services deemed equivalent are available from Mexican sources. As long as purchasers of professional services know what they want, on the basis of proposals submitted, they can clearly evaluate what is being purchased and at what cost.

### Person-to-Person Transfer

As reported by Mexican sources, a critical element in the successful transfer of Bechtel's know-how technology is the human interface; technology cannot be transferred through books, manuals, or computer programs alone, no matter how useful they may be in themselves. Successful technology transfer requires an understanding of the "why's" behind rules, procedures, and systems. This understanding permits consideration of alternative action in unusual or changed conditions. The result is a living technological skill rather than rote knowledge.

In talking with clients of Bechtel's technology transfer programs in Mexico, the selection of people who transfer the technology as well as of those who receive it is critical. For the transferor, the key points cited as characterizing the successful consultant were:

- *Family:* a resident spouse and family supportive of the consultant's mission, sympathetic and adaptable to host nation conditions
- *Attitude:* a "people" orientation in addition to high technical professionalism
- *Patience:* a willingness to accept, where necessary, a tempo and approach different from that to which they may have been accustomed

- *Sensitivity:* a political sensitivity to the environment into which new technology by management system is being introduced.

## Culture Shock

Earlier, the essential nature of Bechtel's technology was defined as the ability to manage complex technical input to produce results of high quality on time and on budget. To do so requires a disciplined, goal-oriented management system. Introduction of such a system into a different society can often present problems of culture shock. In many societies, elements of the system are alien concepts, such as fixed working patterns, a faster-paced life-style, formation of working teams with shared responsibilities, checking on someone else's work, and delegation of responsibility and authority.

Elements of culture shock have been experienced in some of Bechtel's assignments in Mexico. In one instance, the CFE project manager said that it took more than six months to feel comfortable with the assignment and the newly introduced management system. Successful recipients of such advanced technology transfer in addition to technical proficiency must be adaptable, have a willingness to accept new ideas, methods, and systems, and have or develop the wisdom to know and acknowledge what they do not know.

For the recipient of technology transfer, some form of preparation is desirable. In several Bechtel assignments in Mexico, there was virtually no advance preparation; as a result, a Bechtel team arrived to begin work with a hastily assembled group of client counterpart staff. Elements of advanced preparation that may lessen the culture shock that often accompanies technology transfer include:

- Early selection of the counterpart staff and thorough familiarization of that staff with the assignment, work methods, and management system to be employed
- Briefing of the counterpart staff by client top management with respect to the significance of the assignment, what is expected to be learned, and the relationship of the technology transfer experience to the staff's long-term career in the client organization

- Training and practice in conversational English and especially in technical terminology, since most of the initial transfer by Bechtel takes place in English.

### Management Responsibility

In both transferor and recipient company or institution, top management must understand and support the technology transfer effort by word and deed for it to succeed. This calls for the assignment of personnel of the highest caliber and professionalism from both sides—the type of individuals who are always in high demand and in short supply. The technology transfer effort must therefore be considered important by business and corporate managers and viewed as a step in career development.

Service technology transfer as practiced by Bechtel is time consuming and personnel intensive. Often, receiving institutions vastly underestimate the number of people that need to be involved in a successful transfer program. In the last analysis, the critical element of success in the program is the person-to-person transfer of mental attitudes and thinking patterns through on-the-job training under experienced supervision. This transfer takes time.

One Bechtel client in Mexico felt that his organization did not take as much advantage from the technology transfer experience as it could have, because not enough of the client's senior professional staff were directly and intensively involved in the transfer program.

## COMMENTS AND OBSERVATIONS

### What Determines the Extent of Technological Transfer?

The level and extent of technological transfer in any specific project depends on a number of interrelated factors: the client's technology transfer objectives; the educational and experience level of the recipient; the nature of the project and Bechtel's role in project execution; the duration of the project; and the complexity of the technology involved.

The long-term utility of the transfer experience depends on two other factors: the attitude or motivation of the recipient; and the ability of the recipient to utilize the knowledge gained in technology transfer on a continuing basis.

### Client Objectives

There can be little effective technology transfer unless a client wants it, has clearly defined objectives for such a program, and has made the necessary plans to implement it. As previously discussed, top management commitment to such an effort by word and deed is critical in its success.

### Education and Experience Level

The education and experience level of the national trainee sets a base level of entry into formalized training programs, on-the-job training, and more sophisticated levels of technology transfer. Figure 4.3 shows how the training that could be involved in a comprehensive industrial/vocational training differs for employees with six levels of entry qualifications. Formalized training programs can aid in improving skill levels and facilitate job mobility. As noted earlier, in many grass roots Bechtel projects, individuals recruited into the original basic construction work force at a site can, in time, become permanent members of the operating staff of the constructed facility—whether copper smelter, LNG plant, or power station.

For more sophisticated levels of technology transfer, there are educational and experience requirements that must be met before any meaningful program can be undertaken. Where these requirements are not met, Bechtel can and has taken on a counseling role for clients to design educational and apprenticeship programs for selected individuals to bring them up to requirement standards; Bechtel then can monitor participants' progress while in such programs. In connection with development projects in the Middle East, for example, Bechtel has selected educational programs and institutions for individuals being groomed for key engineering and management positions in new companies or government operating agencies. Often these several-year educational programs are supplemented by actual work experience in U.S. or foreign architect-engineering companies or with operating companies. Once these requirements have been met, engineering and managerial on-the-job training programs can be carried out.

### Nature of the Project and Bechtel's Role

As noted earlier, a project contracted for on a turnkey basis usually offers limited opportunities for advanced technology transfer in EPC management services despite the promises of the contractor. Most of the design and engineering work is done offshore, leaving only the actual construction of the project in the host country. In general, such projects are limited to training in construction skills. Training a local work force in construction skills has great economic justification in almost all projects, and has been a standard feature of Bechtel international work since the beginning.

The next level of skills in ease of transfer is composed of O & M skills. Increasingly, clients are explicitly contracting for such formalized training and on-the-job experience.

More sophisticated technology transfer in connection with project work is contracted for on a different basis. It is more time consuming, is more costly because of the training investment in personnel, and requires the commitment of highly trained client personnel who often are in short supply. Generally, advanced technology transfer requires more than one project to be effective.

The opportunities for and effectiveness of technology transfer are greatest where Bechtel has responsibility for project management. Technology transfer is a production task, and time pressure is a vital part of the on-the-job transfer process.

### Duration of the Project

In general, the longer a project, the more opportunity for technology transfer and growth in technical skills.

For example, pipeline projects generally are of limited duration and move rapidly across geographic areas. In these projects even simple construction skills must become specialized. Individual workers are trained in only one aspect of an operation, with little opportunity available to rotate personnel or skills.

At the other end of the spectrum, the example of the Caracas Metro shows how the development of this rapid transit system in phases over a 30-year period has given the opportunity to move it from a project originally designed and engineered by foreigners to one designed, engineered, contracted for, and supervised by Venezuelans. Bechtel's own relationship to the project has changed from that of joint venture participant to that of a partner with a local engineering company that is increasingly assuming more responsibility and control.

**Complexity of the Technology**

The complexity of technology to be transferred determines the required levels of education and experience on the part of the recipients, the number of people that need to be involved, and the length of time required to effect transfer.

For example, in a nuclear power plant project, more than 75 professional disciplines are involved, and in multiproject technology transfer programs as many as 1,300 national engineers can be involved. The transfer of such technology is time consuming and complex. Even one nuclear power project, stretching over ten years, does not provide enough opportunities for moving a qualified trainee through all its different phases of design, engineering, construction, and start-up. A long-term involvement in a series of nuclear power projects is needed so that trainees can assume higher levels of responsibility and control of successive projects using similar procedures and methods.

**Motivation**

Technology transfer, especially in advanced programs, is a person-to-person transfer of technical and managerial experience, attitudes, and viewpoints. It depends for its success on a favorable personal chemistry between transferor and recipient: Both need to be highly motivated for it to work. An eagerness to learn in the recipient breeds enthusiasm in the transferor and vice versa.

For Bechtel professionals involved in advanced technology transfer programs, employees' careers depend on their know-how transfer quality and depth as well as on their performance on project work.

On the other side, the transfer recipient must want to learn, have a desire to excel, and more importantly, be willing to work hard. Although this list of virtues may sound platitudinous, experience shows that such qualities are essential to successful technology transfer.

### Ability to Utilize the Knowledge Gained

The long-term success of any technology transfer program depends on the opportunity for the recipient to continue to practice what he or she has learned to do. That is why clearly thought out objectives for such a program are important. There is not much point in spending the time and money to transfer advanced engineering technology to an individual if the sum total of his or her application horizon is the design of only one or two installations. For acquired skills to grow and develop, there is need for continued education and increasingly challenging application. Without such a challenge with their employers or in their native countries, well-trained individuals will seek out the opportunities wherever they may be for them to develop and achieve. This may be at least one explanation for the "brain drain" from some developing countries.

### Why Does Bechtel Do It?

Since the ultimate goal of technology transfer in advanced Bechtel programs is to enable clients successfully to carry out highly complex projects independently, why then does Bechtel do it? Are they not working themselves out of a job? There are a number of reasons why Bechtel does carry out technology transfer programs:

- *Competition:* In most areas of their professional practice, if Bechtel does not provide the service requested, it is more than likely that others will.
- *Profitability:* Technical transfer programs are not "loss leaders"; Bechtel expects such programs to cover their full operating costs and produce at least normal margins of profit on the workers employed.
- *Dynamic technology:* Although Bechtel does not hold back technology or know-how in its transfer programs, the fact is that most of the technologies they practice are evolving, new processes are being developed, and with each new project more experience is gained. Standards change; operating procedures can change; engineering design modifications constantly occur; problems are being solved and that experience is constantly being fed back into design and construction practice. To this extent, the need for some Bechtel services is not eliminated; rather, the nature of the service it provides to clients changes.
- *New products:* Bechtel strives to maintain itself on the cutting edge of new technology, which gives rise to new services it can provide to clients. This is the main thrust of its Research and Engineering Organization. Work in oil recovery from tar sands, nuclear waste disposal, synthetic oil products derived from coal, and the basic redesign of the conventional fossil-fueled thermal power station to eliminate environmental pollution are some areas that provide the basis for new professional services for Bechtel.
- *New markets:* Further, the market for the technology transfer and engineering and construction management services is itself always evolving. The potential for these Bechtel services in developing nations in many parts of the world is great.
- *Make or buy:* There is always a market for temporary professional services. The possibility of shortening the time needed for the implementation of projects is one reason for obtaining specialized skills; technical expertise needed only occasionally is another.

The service technology transfer as practiced by Bechtel has many characteristics in common with the consulting profession. Individual assignments or projects have limited duration, and the function of the best practitioners in the profession is to work themselves "out of a job." Teaching and training clients to solve problems are an inherent part of many assignments.

# Index

# About the Authors

**Bernard Wasow** received a B.A. from Reed College and an M.A. and Ph.D. from Stanford University. He is Associate Professor and Director of Undergraduate Studies in Economics at New York University specializing in the fields of Economic Development, Economic Theory, and International Economics. He is the author of numerous articles, books, and monographs, including *Dependent Growth in a Capital Importing Economy: The Case of Puerto Rico,* (1978), *The Real Interest Rate, Foreign Capital and Domestic Saving: The Case of Korea,* (1980), *Saving and Dependence with Externally Financed Growth,* (1979), *Labor Force Allocation in Underdeveloped Market Economies,* (1973), and *A Note on Wage/Exchange Rate Policy in an Open Underdeveloped Economy,* (1970). He is coauthor of *Public Sector Involvement in the Insurance Industry: Implications for Economic Development,* (1979) and *The Insurance Industry in Economic Development,* (1982). In addition, he has done consulting work for numerous corporations and organizations, including Westinghouse Corporation, Schering-Plough Corporation, and the Hudson Institute.

**John C. Stephenson** was an independent management consultant specializing in corporate strategic planning, technology transfer, and international economic development programs. He participated in a wide range of marketing, feasibility, diversification, and long range planning studies in the United States and abroad. Prior to this he spent over 25 years as a Senior Staff Associate of Arthur D. Little, Inc. during which time he served in a number of capacities, including responsibility for residential economic development offices in Africa, Asia, and the Middle East. He was a graduate of the Carnegie Mellon University with a B.S. in Chemical Engineering, and Harvard Graduate School of Business Administration.

**Nancy Sherwood Truitt** has been involved in issues of economic and social development in the Third World as a business consultant, foundation executive, and currently as Program Director of the Fund for Multinational Management Education and Executive Director of the U.S. Business Committee on Jamaica. Among her publications are: *Transfer of Technology: An International Issue,* (Ed.), *Opinion Leaders and Private Investment: An Attitude Survey in Chile and Venezuela,* co-authored with David H. Blake, and "The Industrial Community in Peru: An Experiment in Profit Sharing and Collective Management." She has a B.A. from Stanford University and an M.B.A. from New York University Graduate School of Business Administration.

**Ronald Kent Shelp** is a leading authority on services in the world economy. He is Chairman of the U.S. government-appointed Industry Sector Advisory Committee on Services and the Coordinating Committee of the Coalition of Service Industries. Mr. Shelp is Vice President of American International Group, a major New York-based multinational insurance company. He has written extensively on services and other subjects, including a book published in 1981 by Praeger, now in its third printing, *Beyond Industrialization: Ascendancy of the Global Service Economy.* He is also a contributor to a forthcoming book, *The Revitalization of America.* Mr. Shelp holds an undergraduate degree from the University of Georgia and a graduate degree from the Johns Hopkins University School of Advanced International Studies.